# GEOMETRY:
## A STRAIGHTFORWARD APPROACH

2nd Edition

**Martin M. Zuckerman**

*The City College of The City University of New York*

ARDSLEY HOUSE PUBLISHERS, INC., NEW YORK

Address orders and editorial
correspondence to:
Ardsley House, Publishers, Inc.
320 Central Park West
New York, NY 10025



ISBN: 1-880157-63-2

Printed in the United States of America

10 9 8 7 6 5

# CONTENTS

In memory of my grandparents
Wolf and Rebecca Zuckerman
Ben and Rose Blazer

**ACKNOWLEDGMENTS**

The author gratefully acknowledges the many helpful comments of Jacob Barshay, Irwin Bernstein, Walter Daum, Phyllis Getzler, Ralph Kopperman, Rochelle Leon, John Miller, and Dan Mosenkis of City College of the City University of New York.

**Other Titles by Martin M. Zuckerman**

Sets and Transfinite Numbers, Macmillan, 1974
Intermediate Algebra: A Straightforward Approach for College Students, W. W. Norton, 1976
Elementary Algebra Without Trumpets or Drums, Allyn & Bacon, 1976
Arithmetic Without Trumpets or Drums, Allyn & Bacon, 1978

# Chapter 1
# PLANE GEOMETRY

## DIAGNOSTIC TEST FOR CHAPTER 1

Perhaps you are already familiar with some of the material in Chapter 1. This test will indicate which sections of Chapter 1 you need to study. The question number refers to the corresponding section in Chapter 1. If you answer *all* parts of the question correctly, you may omit the section. But *if any part of your answer is wrong, you should study the section.* When you have completed the test, turn to page 165 for the answers.

1.1 Length: American and Metric Systems

Convert from one unit to another, as indicated.

(a) 7 feet into inches

(b) 6 centimeters into meters

(c) 40 kilometers into miles (Express your result to the nearest mile.)

1.2 Rectangles

(a) Find the perimeter of a rectangle of length 5 centimeters and width 2 centimeters.

(b) The perimeter of a square is 100 inches. Find the length of a side.

(c) Find the area of a rectangle with base 7 inches and height 5 inches.

1.3 Triangles, Parallelograms, Trapezoids

(a) Find the area of a triangle with base 6 centimeters and height 5 centimeters.

(b) Find the area of a parallelogram with base 4 feet and height 1 foot.

(c) Find the area of a trapezoid with bases 10 centimeters and 30 centimeters and height 15 centimeters.

1.4 Circles

Express the following areas in terms of $\pi$.

(a) Find the area of a circle with diameter 6 centimeters.

(b) Find the area of a circle with circumference $12\pi$ inches.

1.5 Similar Figures

(a) One rectangle has length 16 inches and width 4 inches. A similar rectangle has length 8 inches. Find its width.

(b) Suppose the ratio of similarity of two squares is 5. If the smaller square has side length 6 inches, find the side length of the larger square.

(c) Suppose the ratio of similarity of two similar parallelograms is 3. If the base of the larger parallelogram is 9 centimeters, find the base of the smaller one.

## 1.1 Length: American and Metric Systems

There are two common systems of measurements: the *American system* and the *metric system.*

### AMERICAN SYSTEM

To measure *length* in the American system, we begin with the basic unit of 1 *inch* (in.). Thus, a 5-inch wire is 5 times as long as a 1-inch wire.

FIGURE 1.1 (a) 1 inch.

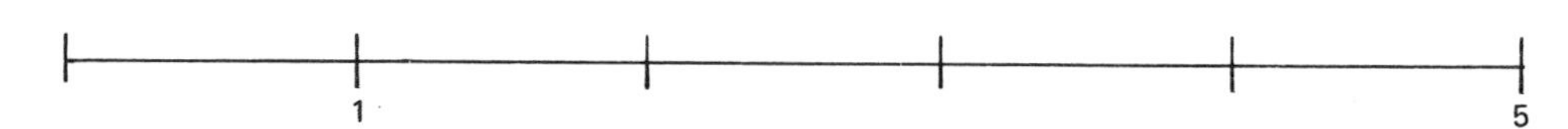


FIGURE 1.1 (b) 5 inches.

To describe larger lengths we use feet (ft.), yards (yd.), and miles (mi.). To change units, use the following:

1 foot = 12 inches
1 yard = 3 feet = 36 inches
1 mile = 1760 yards = 5280 feet

To convert from a *larger* unit *to* a *smaller* unit, *multiply*:

feet into inches, multiply by 12;
yards into feet, multiply by 3;
yards into inches, multiply by 36;
miles into yards, multiply by 1760;
miles into feet, multiply by 5280.

Then change units, as indicated.

EXAMPLE 1. (a) Convert 5 feet into inches.

(b) Convert 3.2 yards into feet.

SOLUTION. (a) 5 ft. = 5 × 12 in.
= 60 in.

(b) 3.2 yd. = (3.2)(3) ft.
= 9.6 ft.

To convert from a *smaller* unit *to* a *larger* unit, *divide*:

inches into feet, divide by 12;
feet into yards, divide by 3;
inches into yards, divide by 36;
yards into miles, divide by 1760;
feet into miles, divide by 5280.

Then change units, as indicated.

EXAMPLE 2. (a) Convert 54 inches into feet.

(b) Convert 144 inches into yards.

SOLUTION. (a) $54 \text{ in.} = \frac{\overset{9}{\cancel{54}}}{\underset{2}{\cancel{12}}} \text{ ft.} = 4\frac{1}{2} \text{ ft.}$

(b) $144 \text{ in.} = \frac{\overset{\overset{4}{\cancel{12}}}{\cancel{144}}}{\underset{\underset{1}{\cancel{3}}}{\cancel{36}}} \text{ yd.} = 4 \text{ yd.}$

## CLASS EXERCISES

Convert from one unit to another, as indicated.

1. 7 feet into inches
2. 100 yards into feet
3. 2¼ yards into inches
4. 1.5 miles into feet
5. 72 inches into feet
6. 72 inches into yards

### METRIC SYSTEM

Powers of 10 play a central role in the metric system. Because 10 is the base of our number system, it is relatively easy to convert units within this system. Most countries use the metric system, and the United States is moving in this direction.

First, you will consider various units of length in the metric system. Then you will compare the American and metric systems.

The basic unit of length in the metric system is the *meter* (m.). 1 meter is approximately 39.37 inches, or a little more than a yard (36 inches).

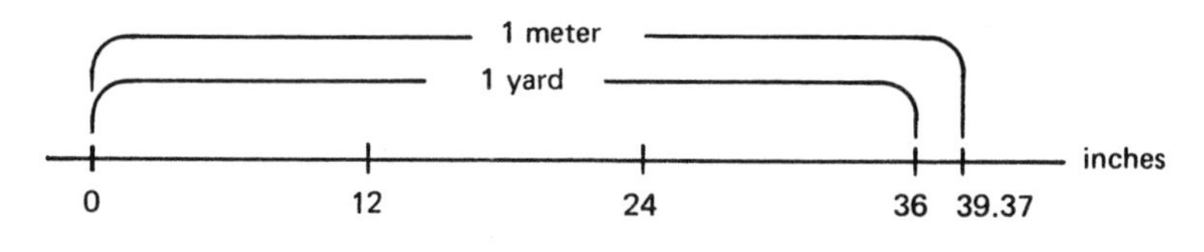


FIGURE 1.2 1 m.≈39.37 in.

1 *kilo*meter (km.) = 1000 meters
1 *hecto*meter (hm.) = 100 meters
1 *deka*meter (dkm.) = 10 meters

There are also units for fractional parts of a meter.

1 *deci*meter (dm.) = $\frac{1}{10}$ meter (or .1 meter)

1 *centi*meter (cm.) = $\frac{1}{100}$ meter (or .01 meter)

1 *milli*meter (mm.) = $\frac{1}{1000}$ meter (or .001 meter)

To convert from one metric unit to another, multiply or divide by the appropriate power of 10. If you convert

from a larger to a smaller unit, multiply;
from a smaller to a larger unit, divide.

Then change units, as called for. For example, to change

kilometers into meters, multiply by 1000;
meters into centimeters, multiply by 100;
millimeters into meters, divide by 1000;
meters into hectometers, divide by 100.

When working with decimals, you can remember how to move the decimal point by thinking of the following chart.

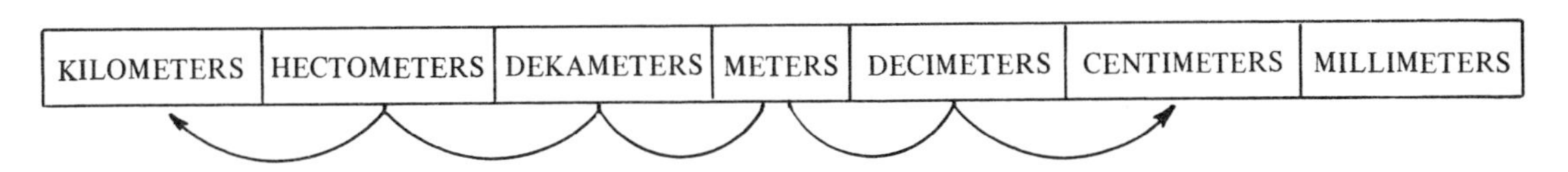


Note that KILOMETERS lies 3 spaces to the *left* of METERS. To convert from meters to kilometers, move the decimal point 3 digits to the *left.*

CENTIMETERS lies 2 spaces to the *right* of METERS. To convert from meters to centimeters, move the decimal point 2 digits to the *right.*

EXAMPLE 3. (a) Convert 5.3 kilometers into meters.

(b) Convert 425 decimeters to meters.

(c) Convert 7.5 meters into centimeters.

SOLUTION. (a) To change from kilometers to meters, multiply by 1000. Here, you must insert two 0's before moving the decimal point 3 digits to the *right.* (Think of the chart.)

5.300 to 5300.

Therefore,

5.3 km. = 5300 m.

(b) To convert decimeters to meters, divide by 10. Thus, move the decimal point 1 digit to the *left.*

425. to 42.5

Therefore,

425 dm. = 42.5 m.

(c) To convert from meters to centimeters, multiply by 100. Here, you must add a zero before moving the decimal point 2 digits to the *right.*

7.50 to 750.

Therefore,

7.5 m. = 750 cm.

## CLASS EXERCISES

Convert from one unit to another, as indicated.

7. 3 kilometers into meters

8. 958 meters into decimeters

9. 5.2 decimeters into meters

10. 950 centimeters into meters

### AMERICAN AND METRIC SYSTEMS

Recall that

1 meter is approximately 39.37 inches.

Write

1 m. ≈ 39.37 in.

To convert inches to meters, divide by 39.37.

```
         .0254
3937./100.0000
       78 74
       21 260
       19 685
        1 5750
        1 5748
             2
```

1 inch ≈ .0254 meter (or 2.54 centimeters)

The comparison is usually with centimeters. Thus,

1 in. ≈ 2.54 cm.

EXAMPLE 4. (a) Convert 12 meters into inches. Express your result to the nearest inch.

(b) Convert 60 inches into centimeters. Express your result to the nearest centimeter.

SOLUTION. (a)

```
  39.37
 ×   12
 472.44
```

To the nearest inch,

12 m. ≈ 472 in.

(b)

```
   2.54
 ×   60
 152.40
```

To the nearest centimeter,

60 in. ≈ 152 cm.

There is also a useful conversion between kilometers and miles.

1 km. ≈ .62 mi.

1 mi. ≈ 1.61 km.

EXAMPLE 5. If you are 58 kilometers from Mexico City, what is the distance to Mexico City in miles?

SOLUTION.

$$\begin{array}{r} 58 \\ \times\ .62 \\ \hline 1\ 16 \\ 34\ 8\phantom{0} \\ \hline 35.96 \end{array}$$

The distance to Mexico City is about 36 miles.

## CLASS EXERCISES

Convert from one unit to another, as indicated. Express your result to the nearest unit.

11. 30 meters into inches
12. 9.2 inches into centimeters
13. 38 kilometers into miles
14. 20 miles into kilometers

## SOLUTIONS TO CLASS EXERCISES

1. 7 ft. = 7 × 12 in. = 84 in.
2. 100 yds. = 100 × 3 ft. = 300 ft.
3. $2\frac{1}{4}$ yds. = $\frac{9}{\not{4}_1} \times \not{36}^{9}$ in. = 81 in.
4. 1.5 mi. = (1.5) 5280 ft. = 7920 ft.
5. 72 in. = $\frac{72}{12}$ ft. = 6 ft.
6. 72 in. = $\frac{72}{36}$ yd. = 2 yd.
7. 3 km. = 3 × 1000 m. = 3000 m.
8. 958 m. = 958 × 10 dm. = 9580 dm.
9. 5.2 dm. = $\frac{5.2}{10}$ m. = .52 m.
10. 950 cm. = $\frac{950}{100}$ m. = 9.5 m.
11. 30 m. ≈ $\underbrace{30 \times 39.37}_{1181.10}$ in.

    To the nearest inch,
    30 m. ≈ 1181 in.
12. 9.2 in. ≈ $\underbrace{9.2 \times 2.54}_{23.368}$ cm.

    To the nearest centimeter,
    9.2 in. ≈ 23 cm.

13. $38 \text{ km.} \approx \underbrace{38 \times .62}_{23.56} \text{ mi.}$

To the nearest mile,
38 km. ≈ 24 mi.

14. $20 \text{ mi.} \approx \underbrace{20 \times 1.6}_{32.2} \text{ km.}$

To the nearest kilometer,
20 mi. ≈ 32 km.

NAME ____________

CLASS ____________

## HOME EXERCISES

ANSWERS

In Exercises 1-24, convert from one unit to another, as indicated.

1. 96 inches into feet
2. 108 inches into yards
3. 18 feet into yards
4. 10,560 feet into miles
5. 12,320 yards into miles
6. 100 inches into feet
7. 32 inches into yards
8. 25 feet into yards
9. 13.5 feet into yards
10. 6 feet into inches
11. 3 miles into feet
12. 3 miles into yards
13. 9 1/3 yards into feet
14. 2 1/2 feet into inches
15. 1/10 mile into feet
16. 1 mile into inches
17. 5 kilometers into meters
18. 7 decimeters into meters
19. 405 centimeters into meters
20. 3826 millimeters into meters
21. 2440 meters into kilometers
22. 8.5 meters into decimeters
23. 3.58 meters into centimeters
24. 1.2 meters into millimeters

In Exercises 25-38, convert from one unit to another, where indicated. Express your result to the nearest unit.

25. 50 meters into inches
26. 23 meters into inches
27. 10 inches into centimeters
28. 1 foot into centimeters
29. 50 kilometers into miles
30. 12 kilometers into miles
31. 30 miles into kilometers
32. 54 miles into kilometers

1. ____________
2. ____________
3. ____________
4. ____________
5. ____________
6. ____________
7. ____________
8. ____________
9. ____________
10. ____________
11. ____________
12. ____________
13. ____________
14. ____________
15. ____________
16. ____________
17. ____________
18. ____________
19. ____________
20. ____________
21. ____________
22. ____________
23. ____________
24. ____________
25. ____________
26. ____________
27. ____________
28. ____________
29. ____________
30. ____________
31. ____________
32. ____________

ANSWERS

33. ____________

34. ____________

35.(a) ____________

(b) ____________

36. ____________

37. ____________

38. ____________

33. The distance between New York City and Boston is 210 miles. Express this distance in kilometers. Use 1 mi. $\approx$ 1.609 km.

34. The distance between Paris and Brussels is 295 kilometers. Express this distance in miles. Use 1 km. $\approx$ .621 mi.

35. A football field is 100 yards long.
    (a) How many inches long is it? (b) How many centimeters long is it?

36. Which is the longest?
    (a) 50 yards (b) 140 feet (c) 45 meters

37. A car gets 31 miles to a gallon. How many gallons does it need to travel 100 kilometers?

38. The center for the Yugoslavian Olympic basketball team is 203.2 cm. tall. What is his height in feet and inches?

## 1.2 Rectangles

### PERIMETER

DEFINITION. A *polygon* is a closed figure in the plane, composed of 3 or more line segments. A 3-sided polygon is called a *triangle* and a 4-sided polygon is called a *quadrilateral.*

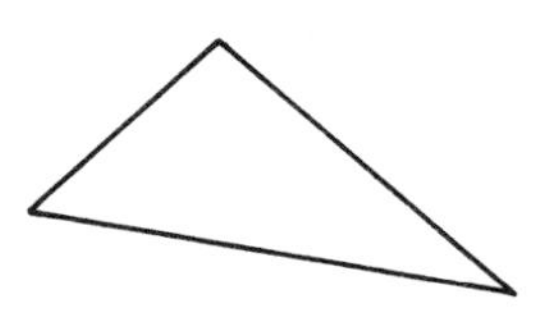

FIGURE 1.3 A triangle.

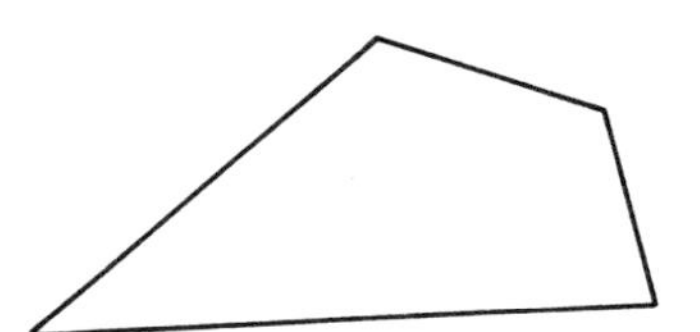

FIGURE 1.4 A quadrilateral.

DEFINITION. The *perimeter* of a polygon is the sum of the lengths of its sides.

The perimeter of a polygon can be thought of as the distance around it.

EXAMPLE 1. Find the perimeter of the polygons in Figure 1.5.

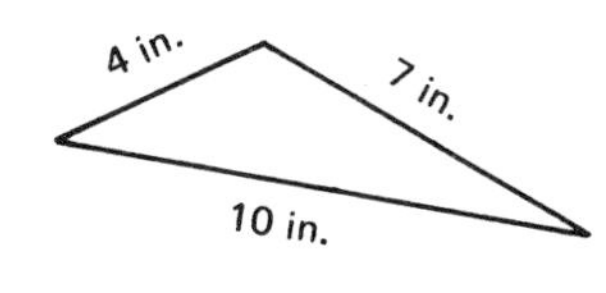


FIGURE 1.5 (a)

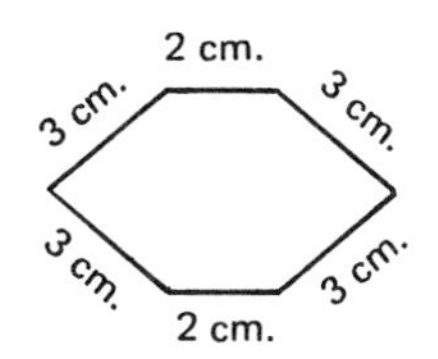


FIGURE 1.5 (b)

SOLUTION. In each case, the expressions to be added involve the *same units.* Add in columns, as indicated.

(a)
4 in.
7 in.
10 in.
21 in.

(b)
2 cm.
3 cm.
3 cm.
2 cm.
3 cm.
3 cm.
16 cm.

## CLASS EXERCISES

Find the perimeter of each polygon.

1.

1 in.
2 in.
1 in.
3 in.

FIGURE 1.6

2.

2 cm.
4 cm.
5 cm.
6 cm.
5 cm.

FIGURE 1.7

3.

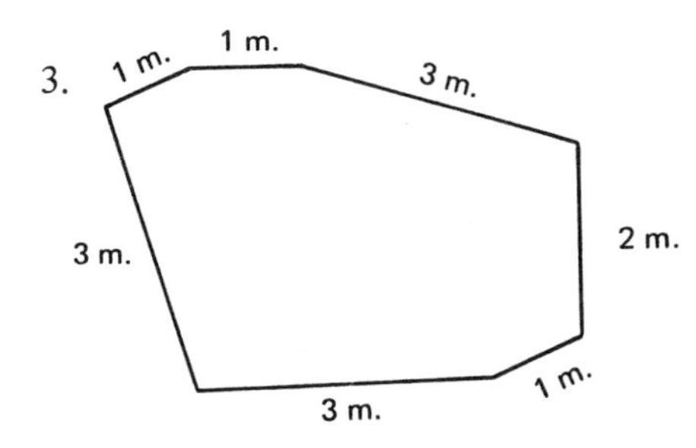


FIGURE 1.8

## ANGLES

*Angles* are formed when lines intersect. The *vertex* (plural: *vertices*) of an angle is the point of intersection of the lines. The vertices of a polygon are the intersections of the line segments of the polygon.

An angle is measured in *degrees*; normally, an angle measures between $0^\circ$ (0 degrees) and $360^\circ$. In Figure 1.9, several angles and their measurements are shown. The $90^\circ$ angle in Figure 1.9 (b) is called a *right angle* and the intersecting lines are said to be *perpendicular.*

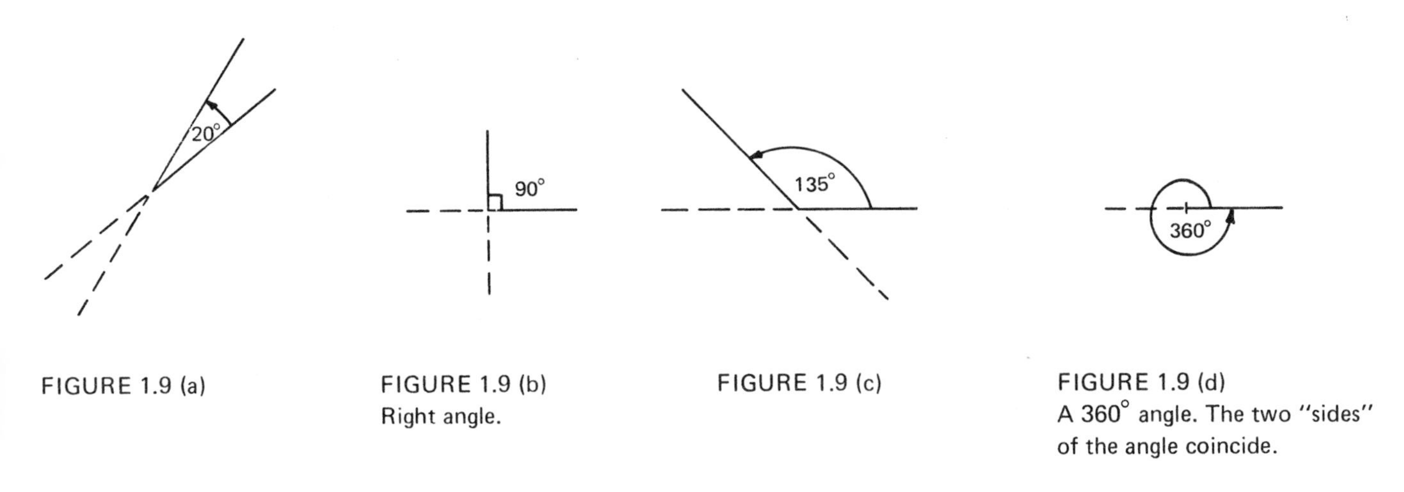


FIGURE 1.9 (a)

FIGURE 1.9 (b)
Right angle.

FIGURE 1.9 (c)

FIGURE 1.9 (d)
A 360° angle. The two "sides" of the angle coincide.

## RECTANGLES

DEFINITION. A *rectangle* is a quadrilateral, or 4-sided polygon, in which each (interior) angle is a right angle.

In Figure 1.10 different quadrilaterals are shown. Only part (a) depicts a rectangle. Note that *the opposite sides of a rectangle are of equal length.*

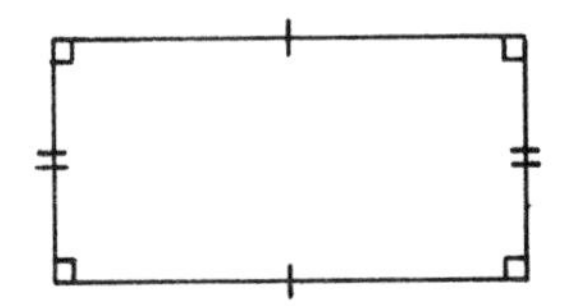

FIGURE 1.10 (a) This quadrilateral is a rectangle. Opposite sides are of equal length and are marked with the same number of bars. Note that each angle is a right angle.

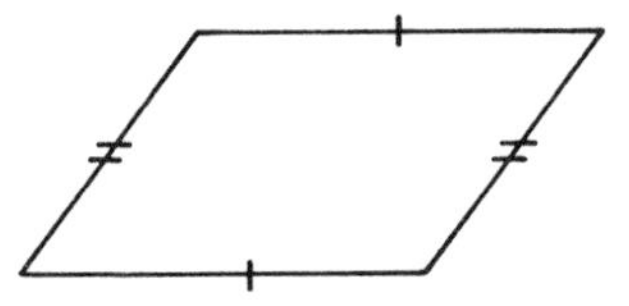

FIGURE 1.10 (b) This quadrilateral (which is called a parallelogram) is not a rectangle, even though opposite sides are of equal length. The angles are not right angles.

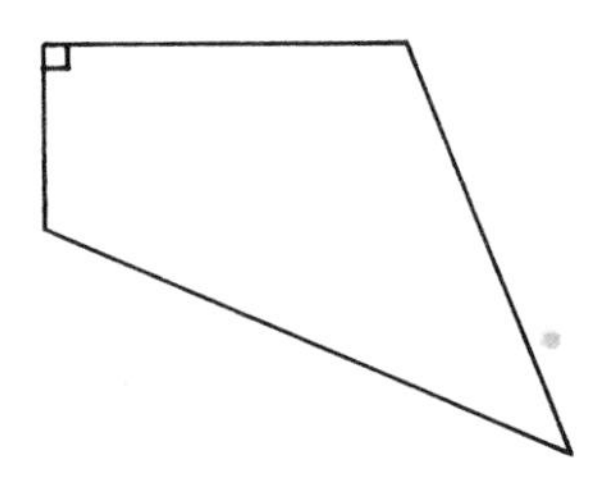

FIGURE 1.10 (c) This quadrilateral is not a rectangle. Only one angle is a right angle.

Note that the perimeter of a rectangle is twice the sum of its length and width.

*Perimeter of rectangle = 2 (length + width)*

Thus, in Figure 1.11,

$$\begin{array}{r} \text{length } 6 \text{ in.} \\ + \text{width } \underline{3 \text{ in.}} \\ 9 \text{ in.} \\ \underline{\times\ 2\quad} \\ 18 \text{ in.} \end{array}$$

When you multiply 9 *inches* by (the number) 2, the product is (9 × 2) *inches.*

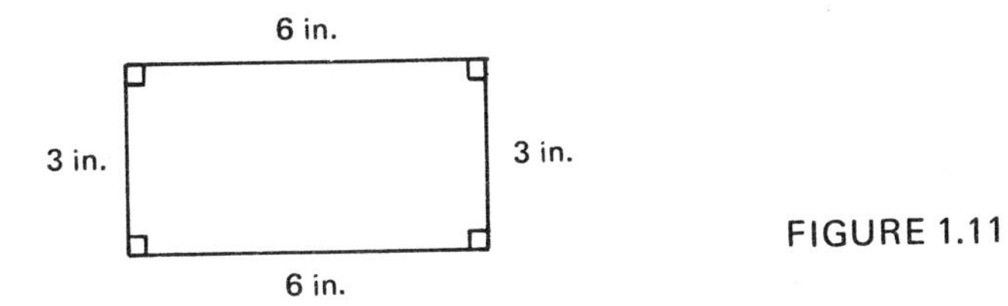


FIGURE 1.11

The area of a rectangle† is the product of length and width.

Area = length × width

In Figure 1.13, the length of the rectangle is 5 inches and the width is 3 inches. The area is:

$$\underbrace{\text{length}} \times \underbrace{\text{width}}$$

5 inches × 3 inches = 15 inches × inches

or 15 *square inches*

†In more precise language, we would speak of the area of a *rectangular region,* or the area enclosed by a rectangle. (See Figure 1.12.) For brevity, we will speak of the area of a rectangle. Similarly, we will refer to the area of a triangle (rather than of a triangular region), the area of a circle (rather than of a circular region), etc.

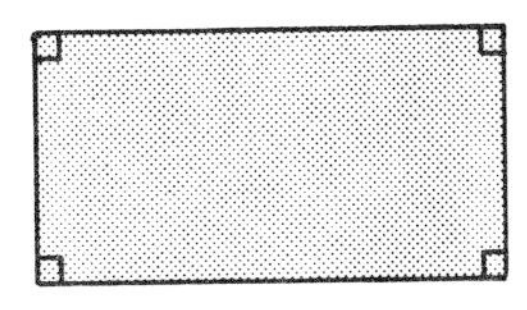

FIGURE 1.12 The rectangle is in black.
The rectangular region is shaded.

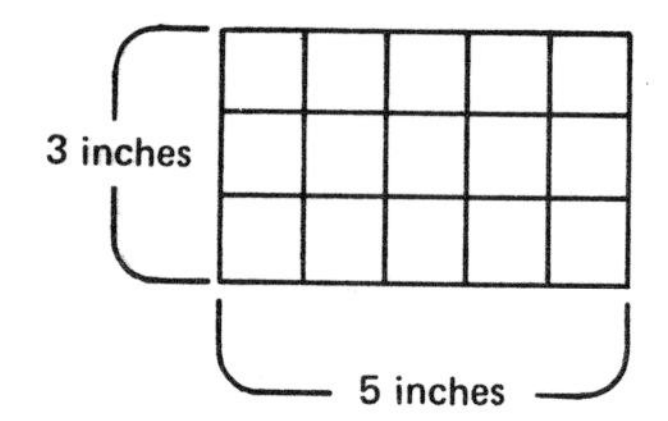


FIGURE 1.13 The area of the rectangle is 15 square inches.

Note that the unit of area is *square inches.*

inches × inches = square inches

Write

in.$^2$ for square inches.

Similarly, write

ft.$^2$ for square feet,
yd.$^2$ for square yards,
mi.$^2$ for square miles,

and in the metric system,

cm.$^2$ for square centimeters,
m.$^2$ for square meters,
km.$^2$ for square kilometers,

etc.

When rectangles are depicted as in Figure 1.14, it is customary to call the horizontal side of each the *base* and the vertical side the *height,* regardless of which side is longer.† Thus,

area of rectangle = base × height

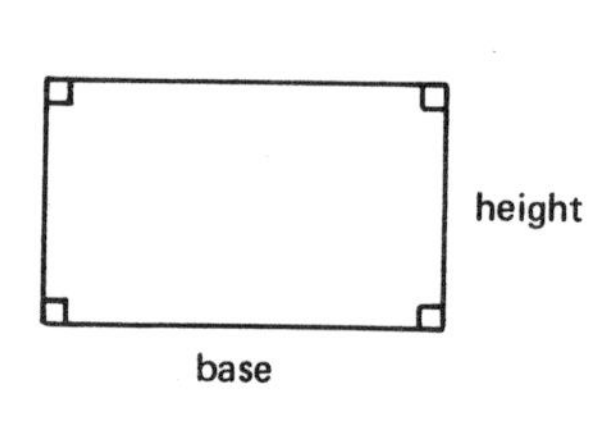

FIGURE 1.14 (a) base = length
height = width

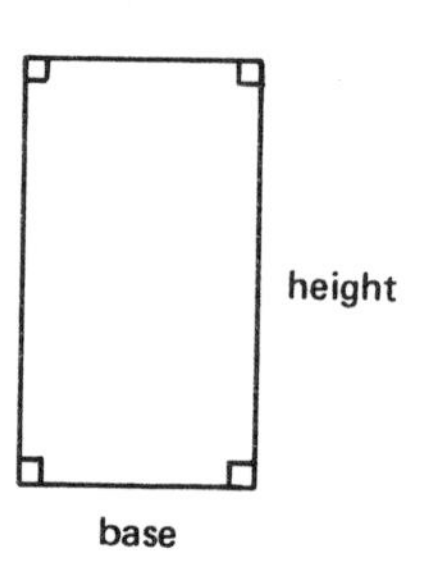


FIGURE 1.14 (b) base = width
height = length

† As is the usual custom, we will use the terms "base" and "height" both for the sides of a rectangle as well as for the lengths of these sides. Similarly, we will use terms such as "radius," "diameter," etc., both for line segments and for the lengths of these line segments.

Let $A$ = area,
$b$ = base,
$h$ = height.

In symbols, the formula for area is:

$$A = b \cdot h$$

EXAMPLE 2. Find the area of the rectangle with base 20 centimeters and height 10 centimeters.

SOLUTION.

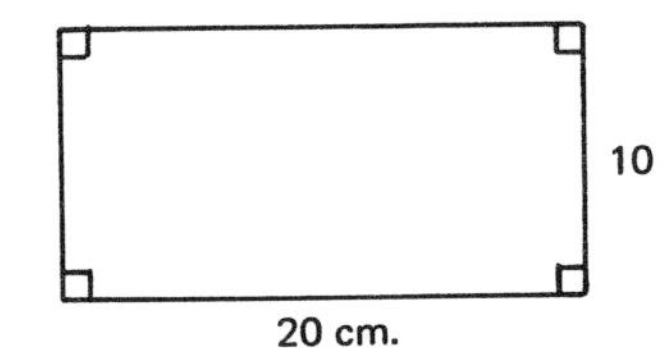


FIGURE 1.15

$$\begin{aligned} A &= b \cdot h \\ &= 20 \text{ cm.} \times 10 \text{ cm.} \\ &= 200 \text{ cm.}^2 \end{aligned}$$

EXAMPLE 3. The perimeter of a rectangle is 36 inches. Its area is 80 square inches. Find the dimensions.

SOLUTION. Let $\ell$ and $w$ be the dimensions of the rectangle (in inches). Then

$$2\ell + 2w = 36$$

and

$$\ell w = 80$$

From the first equation,

$$\ell + w = 18,$$

$$w = 18 - \ell$$

The second equation,

$$\ell w = 80,$$

becomes

$$\ell\,(18 - \ell) = 80,$$

$$18\ell - \ell^2 = 80,$$

or

$$\ell^2 - 18\ell + 80 = 0 \qquad \text{Factor the left side.}$$

$$(\ell - 10)(\ell - 8) = 0$$

For the product to be 0, one of the factors must be 0.

| | |
|---|---|
| $\ell - 10 = 0$ | $\ell - 8 = 0$ |
| $\ell = 10$ | $\ell = 8$ |
| $w = 18 - \ell$ | $w = 18 - \ell$ |
| $w = 8$ | $w = 10$ |

The rectangle is 8 inches by 10 inches.

## CLASS EXERCISES

In Exercises 4 and 5, find the area of each rectangle.

4. base 8 inches, height 3 inches

5. base 5 meters, height 12 meters

6. Find the height of a rectangle with base 4 inches and area 20 square inches.

7. The perimeter of a rectangle is 64 feet and the area is 240 square feet. Find the dimensions of the rectangle.

## SQUARES

**DEFINITION.** A *square* is a rectangle in which each side has the same length.

In Figure 1.16, the length of each side of the square is 6 inches. The perimeter is 24 inches.

FIGURE 1.16

$$\begin{array}{r} 6 \text{ in.} \\ 6 \text{ in.} \\ 6 \text{ in.} \\ \underline{6 \text{ in.}} \\ 24 \text{ in.} \end{array}$$

The length and width of a square are equal. Therefore,

$$\text{Perimeter} = 2\,(\text{length} + \overset{\text{length}}{\cancel{\text{width}}})$$

$$= 2\,(2 \times \text{length})$$

$$= 4 \times \text{length}$$

Thus, *the perimeter of a square is 4 times the length of a side.*

The *area of a square* is given by

$$A = s^2,$$

where s is the length of a side. Thus, for a square of side length 6 inches, the area is 36 square inches.

EXAMPLE 4. Find (a) the perimeter and (b) the area of a square of side length 16 centimeters.

SOLUTION.

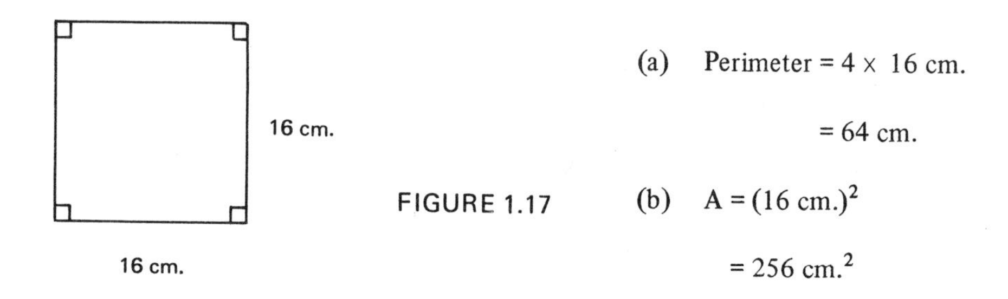


FIGURE 1.17

(a) $\text{Perimeter} = 4 \times 16 \text{ cm.}$

$= 64 \text{ cm.}$

(b) $A = (16 \text{ cm.})^2$

$= 256 \text{ cm.}^2$

## CLASS EXERCISES

8. Find (a) the perimeter and (b) the area of a square of side length 5 inches.

## SOLUTIONS TO CLASS EXERCISES

1.
1 in.
1 in.
3 in.
2 in.
7 in.

2.
2 cm.
5 cm.
5 cm.
6 cm.
4 cm.
22 cm.

3.
1 m.
3 m.
2 m.
1 m.
3 m.
3 m.
1 m.
14 m.

4. $A = b \cdot h$
$= 8 \text{ in.} \times 3 \text{ in.}$
$= 24 \text{ in.}^2$

5. $A = b \cdot h$
$= 5 \text{ m.} \times 12 \text{ m.}$
$= 60 \text{ m.}^2$

6. Area = base × height

$20 \text{ in.}^2 = 4 \text{ in.} \times h$

$$\frac{20 \text{ in.}^2}{4 \text{ in.}} = h$$

To divide expressions with units:

(1) Divide the numbers.

(2) Divide the units.

$$\frac{\overset{5 \text{ in.}}{\cancel{20}\,\cancel{\text{in.}^2}}}{\underset{1}{\cancel{4}\,\cancel{\text{in.}}}} = h$$

$5 \text{ in.} = h$

7. Let $\ell$ and $w$ be the dimensions of the rectangle (in feet).

Then
$2\ell + 2w = 64$ The perimeter is 64 feet.
$\ell + w = 32$
and $w = 32 - \ell$

Also, $\ell w = 240$ The area is 240 square feet.
$\ell(32 - \ell) = 240$
$32\ell - \ell^2 = 240$
$\ell^2 - 32\ell - 240 = 0$

$(\ell - 20)(\ell - 12) = 0$

$\ell - 20 = 0$
$\ell \quad = 20$
$w \quad = 32 - \ell$
$w \quad = 12$

$\ell - 12 = 0$
$\ell \quad = 12$
$w \quad = 32 - \ell$
$w \quad = 20$

The rectangle is 20 feet by 12 feet.

(a) Perimeter = $4 \times 5$ in. = 20 in.

(b) Area = $(5 \text{ in.})^2 = 25 \text{ in.}^2$

NAME ______

CLASS ______

## HOME EXERCISES

In Exercises 1-5, find the perimeter of each polygon.

1.

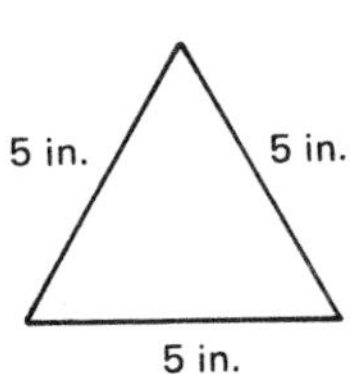


FIGURE 1.18

2.

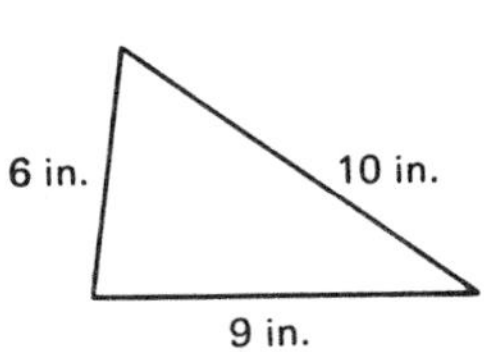


FIGURE 1.19

3.

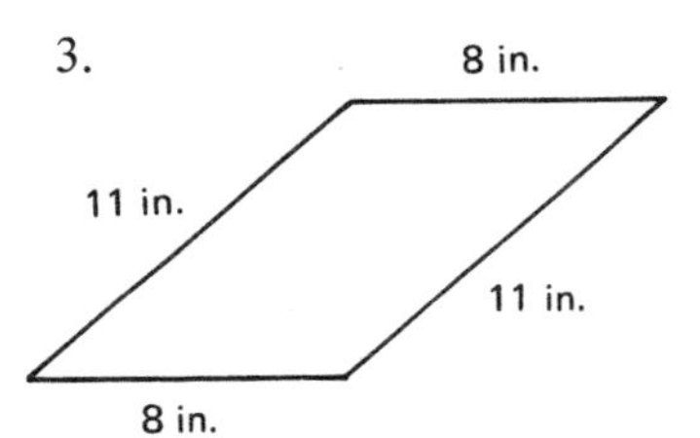


FIGURE 1.20

4.

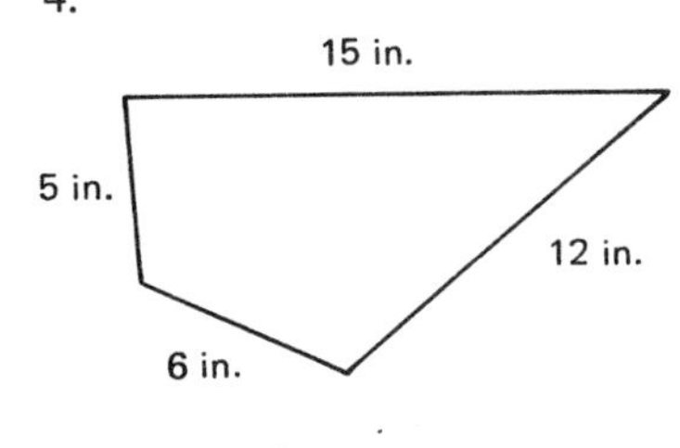


FIGURE 1.21

5.

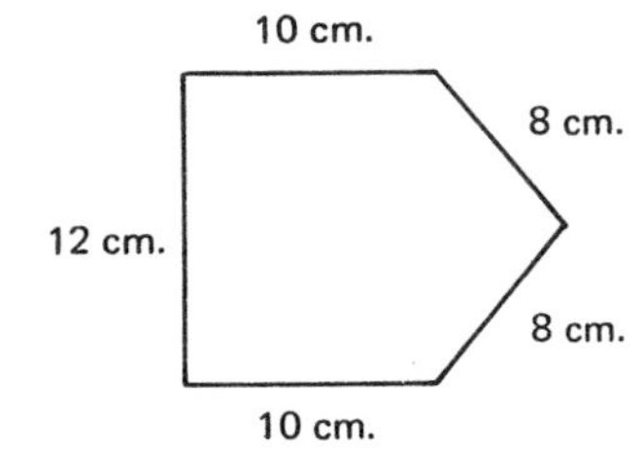


FIGURE 1.22

In Exercises 6-10, find the perimeter of each rectangle.

6. length = 9 inches, width = 2 inches

7. length = 10 inches, width = 1 inch

8. length = 6 centimeters, width = 5 centimeters

9. length = 7 feet, width = 3 feet

10. length = 4 miles, width = 1/2 mile

In Exercises 11-14, find the perimeter of the square whose side length is as indicated.

11. 1 foot

12. 7 centimeters

13. 1.5 miles

14. 2/3 mile

15. The perimeter of a square is 40 inches. Find the length of a side.

16. The perimeter of a square is 2 feet. Find the length of a side.

17. How much rope is needed to enclose a rectangular field 90 feet by 50 feet?

18. How much rope is needed to enclose a rectangular field 90 feet by 50 feet and to divide it in two, (a) lengthwise? (b) widthwise?

ANSWERS

1. ______
2. ______
3. ______
4. ______
5. ______
6. ______
7. ______
8. ______
9. ______
10. ______
11. ______
12. ______
13. ______
14. ______
15. ______
16. ______
17. ______
18.(a) ______

(b) ______

ANSWERS

19. ____________
20. ____________
21. ____________
22. ____________
23. ____________
24. ____________
25. ____________
26. ____________
27. ____________
28. ____________
29. ____________
30. ____________
31. ____________
32. ____________
33. ____________
34. ____________
35. ____________
36. ____________

19. A baseball diamond is a square with side length 90 feet. What is the minimum distance a player must run when he hits a home run?

20. In Figure 1.23, a square is cut from each corner of a 14-inch by 10-inch rectangle. If a side of each square is 2 inches, find the perimeter of the figure.

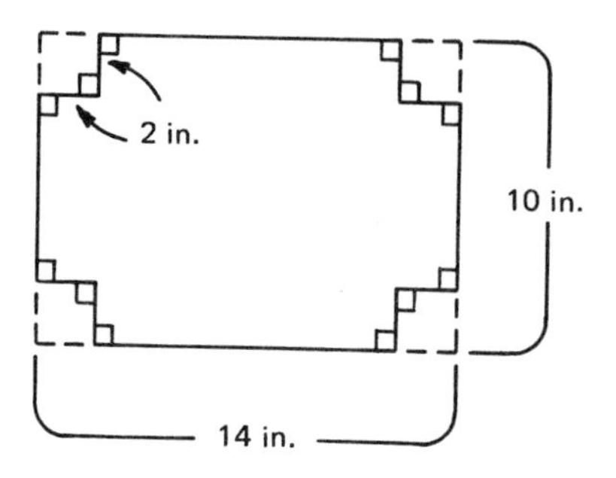


FIGURE 1.23

In Exercises 21-26, find the area of each rectangle.

21. base 4 inches, height 3 inches

22. base 5 feet, height 7 feet

23. base 25 centimeters, height 12 centimeters

24. base 9 inches, height 8 inches

25. base 3/4 inch, height 2 inches

26. base 2/3 inch, height 9/10 inch

In Exercises 27-29, find the base of each rectangle.

27. height 5 inches, area 20 square inches

28. height 2 feet, area 4 square feet

29. height 25 centimeters, area 75 square centimeters

In Exercises 30-32, find the height of each rectangle.

30. base 2 inches, area 20 square inches

31. base 4 yards, area 24 square yards

32. base 9 centimeters, area 108 square centimeters

33. Find the area of a square of side length 9 inches.

34. Find the length of a side of a square whose area is 49 square yards.

35. How many square feet of material are needed to make a tablecloth that is 6 feet by 4 feet?

36. A piece of typing paper measures 8 1/2 inches by 11 inches. What is its area?

NAME

CLASS

ANSWERS

37. A football field measures 100 yards by 53.333 yards. To the nearest square yard, what is its area?

38. One square has side length twice that of another square. If the smaller square has area A, express the area of the larger square in terms of A.

39. The floor of a room measures 20 feet by 14 feet. How much does it cost to varnish the floor at 10 cents per square foot?

40. Suppose the ceiling of the room in Exercise 39 is 9 feet high. How much does it cost to paint the 4 sides of the room and the ceiling at 8 cents per square foot?

41. The perimeter of a rectangle is 70 inches and the area is 300 square inches. Find the dimensions of the rectangle.

42. The area of a garden is 1200 square feet. If the length of the garden were increased by 5 feet and the width decreased by 5 feet, the area would be decreased by 75 square feet. Find the dimensions of the garden.

43. The area of a Chinese rug is 360 square feet. A Persian rug, which is 4 feet longer but 3 feet narrower, has the same area. Find the dimensions of the Chinese rug.

In Exercises 44-46, find the areas of the indicated figures, which are composed of rectangles.

44.

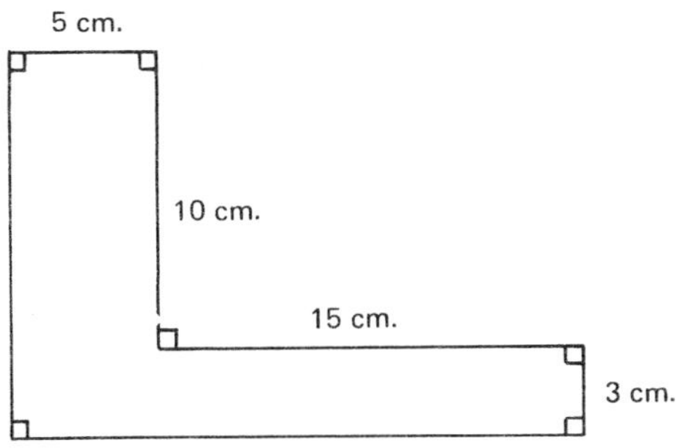

5 cm.

10 cm.

15 cm.

3 cm.

FIGURE 1.24 (a)

45.

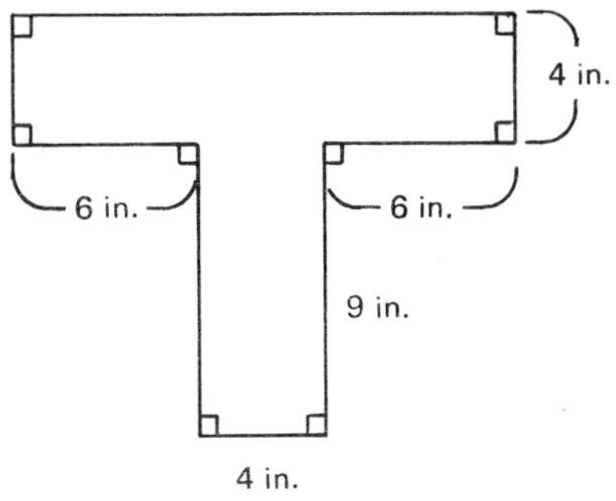

4 in.

6 in.

6 in.

9 in.

4 in.

FIGURE 1.24 (b)

37. ______

38. ______

39. ______

40. ______

41. ______

42. ______

43. ______

44. ______

45. ______

ANSWERS

46. ____________________

46.

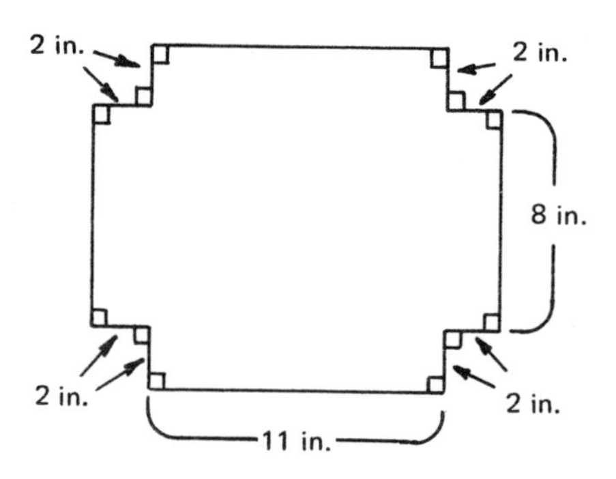


FIGURE 1.24 (c)

## 1.3 Triangles, Parallelograms, Trapezoids

### TRIANGLES

Recall that a triangle is a 3-sided polygon. The line segment from any vertex drawn perpendicular to the opposite side is called a *height* of the triangle. The corresponding perpendicular side of the triangle is called a *base.*

Several triangles are shown in Figure 1.25. A base, b, and corresponding height, h, are indicated in each case.

The triangle in Figure 1.25 (a) is known as a *right triangle.* One of its angles is a right angle.

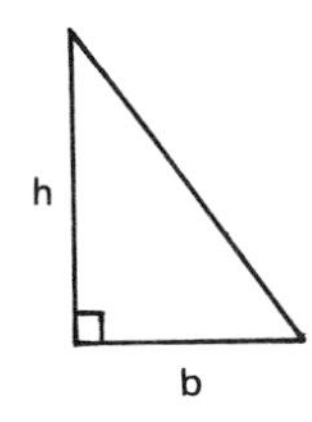


FIGURE 1.25 (a)
Right triangle.

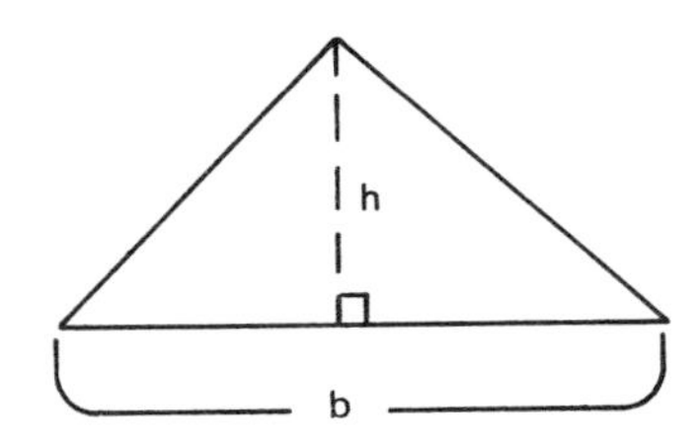


FIGURE 1.25 (b)

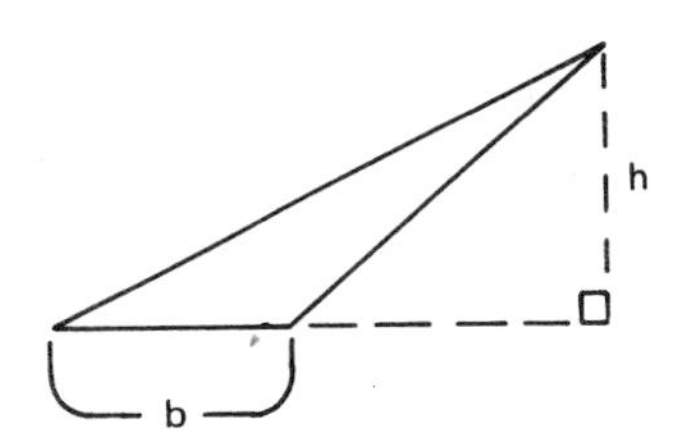


FIGURE 1.25 (c)

The formula for the area of a triangle can be obtained from the formula for the area of a rectangle. A right triangle can be thought of as "half of a rectangle." Therefore, its area is half that of the corresponding rectangle.

$$\text{Area of triangle} = \frac{1}{2}\,\text{Area of rectangle}$$
$$= \frac{1}{2}\,b \cdot h$$
$$= \frac{b \cdot h}{2}$$

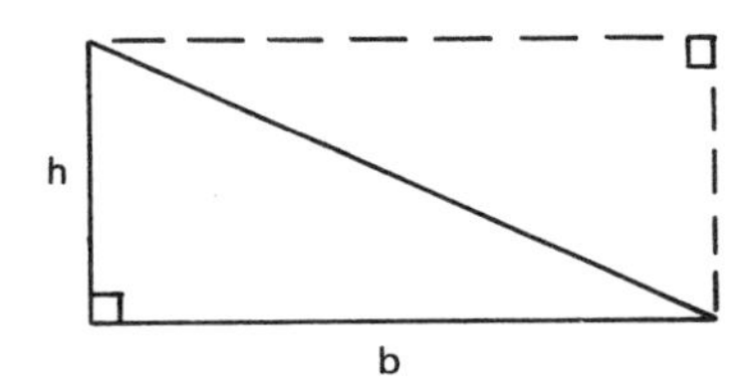


FIGURE 1.26

EXAMPLE 1. Find the area of a right triangle with base 6 inches and height 5 inches.

SOLUTION.

$$\text{Area} = \frac{b \cdot h}{2}$$

$$= \frac{\overset{3\text{ in.}}{\cancel{6\text{ in.}}} \times 5\text{ in.}}{\underset{1}{\cancel{2}}}$$

$$= 15\text{ in.}^2$$

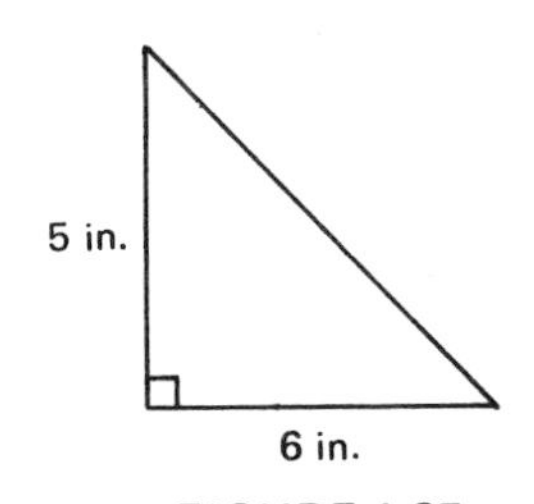


FIGURE 1.27

For other types of triangles, the formula for area is the same. For example, consider the triangle in Figure 1.28 (a), with base 10 inches and height 5 inches. Its area is the sum of the areas of the two *right* triangles in Figure 1.28 (b), with bases 4 inches and 6 inches and with common height 5 inches.

Note that
$$A_1 = \frac{\overset{2 \text{ in.}}{\cancel{4 \text{ in.}}} \times 5 \text{ in.}}{\underset{1}{\cancel{2}}} = 10 \text{ in.}^2$$

and
$$A_2 = 15 \text{ in.}^2$$
(as in Example 1).

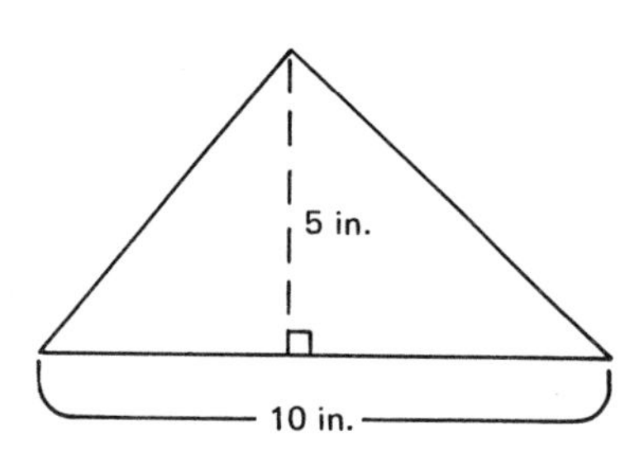


FIGURE 1.28 (a)

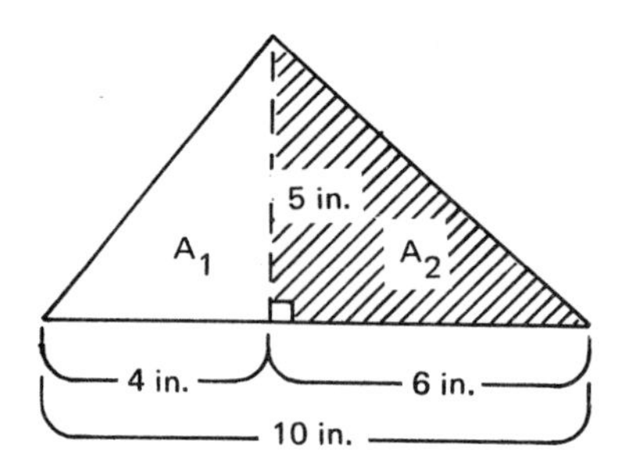


FIGURE 1.28 (b)

Let A be the area of the triangle in Figure 1.28 (a).

$$A = A_1 + A_2$$
$$= 10 \text{ in.}^2 + 15 \text{ in.}^2 \text{ [See Fig. 1.28 (b).]}$$
$$= 25 \text{ in.}^2$$

Observe that
$$25 \text{ in.}^2 = \frac{50 \text{ in.}^2}{2}$$
$$= \frac{10 \text{ in.} \times 5 \text{ in.}}{2},$$

that is,
$$\frac{\text{base} \times \text{height}}{2}$$
[See Figure 1.28 (a).]

To sum up, *for any triangle,*

*Area = ½ base · height,*

or in symbols,

$$A = \frac{1}{2}\, b \cdot h$$
$$A = \frac{b \cdot h}{2}$$

Actually, the area of a triangle can be determined by using *any* side as base, b, together with the corresponding height, h. Thus, in Figure 1.28 (c).

$$A = \frac{b_1 \cdot h_1}{2} = \frac{b_2 \cdot h_2}{2} = \frac{b_3 \cdot h_3}{2}$$

Note, further, that the three heights intersect at a common point.

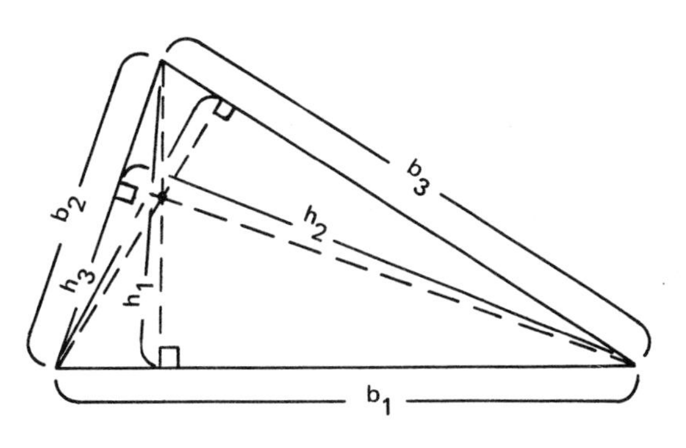


FIGURE 1.28 (c)

EXAMPLE 2. Find the area of the triangle in Figure 1.29.

SOLUTION.

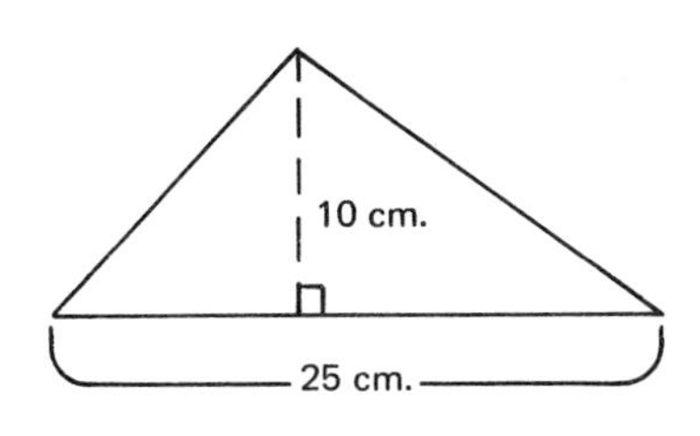


FIGURE 1.29

$$A = \frac{b \cdot h}{2} = \frac{25 \text{ cm.} \times \overset{5 \text{ cm.}}{\cancel{10 \text{ cm.}}}}{\underset{1}{\cancel{2}}}$$

$$= 125 \text{ cm.}^2$$

## CLASS EXERCISES

In Exercises 1-3, find the area of each triangle.

1. base 8 inches, height 4 inches

2.

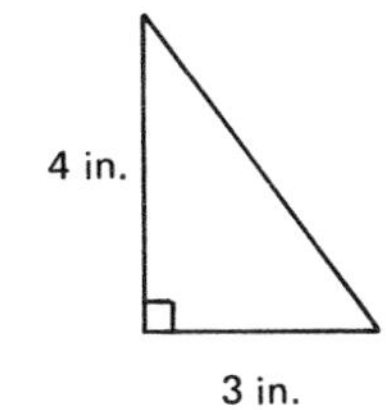


FIGURE 1.30

3.

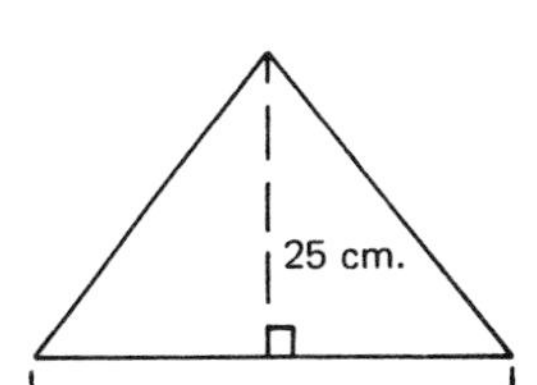


FIGURE 1.31

4. In Figure 1.32, note that there are two right triangles: the smaller one with base 16 cm. and the larger one with base 24 cm. (= 16 cm. + 8 cm.)

   (a) Find the area of the smaller right triangle.

   (b) Find the area of the larger right triangle.

   (c) Find the area of the third triangle (darkened) and show that this is the area of the larger right triangle minus the area of the smaller right triangle.

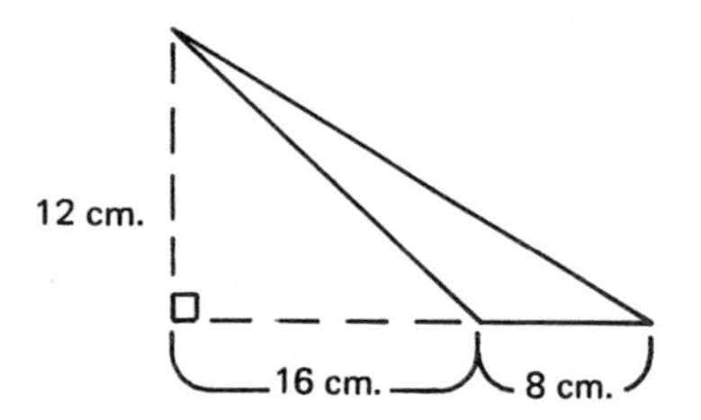


FIGURE 1.32

## PYTHAGOREAN THEOREM

In a right triangle, the side opposite the right angle is known as the *hypotenuse.*

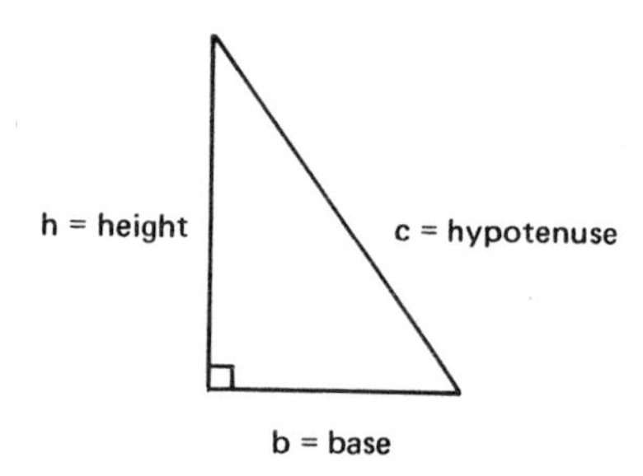


FIGURE 1.33 (a)

Let

$$b = \text{base}$$
$$h = \text{height}$$
$$c = \text{hypotenuse.}$$

The *Pythagorean Theorem* states that

$$b^2 + h^2 = c^2.$$

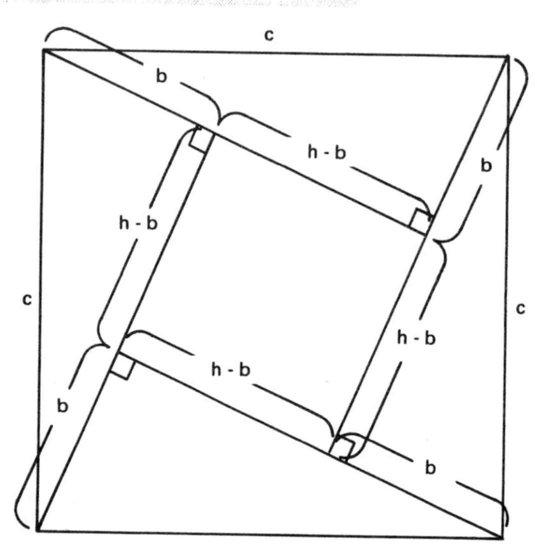


FIGURE 1.33 (b)

To see this, consider Figure 1.33 (b). A square of side length c is divided up to form 4 right triangles and a smaller square. For each right triangle, the base is b, the height is b + (h – b), or h, and the hypotenuse is c.†

Area of 4 right triangles + area of small square = area of large square

$$4 \cdot \frac{b \cdot h}{2} + (h-b)^2 = c^2$$

$$\frac{\overset{2}{\not{4}}bh}{\underset{1}{\not{2}}} + h^2 - 2bh + b^2 = c^2$$

$$h^2 + b^2 = c^2$$

or

$$b^2 + h^2 = c^2$$

EXAMPLE 3. Find the hypotenuse of the right triangle with base 3 inches and height 4 inches.

SOLUTION. Let b, h, and c be as above. Then c, the (*length* of the) hypotenuse is positive.

$$b^2 + h^2 = c^2$$

$$3^2 + 4^2 = c^2$$

$$9 + 16 = c^2$$

$$25 = c^2$$

$$\sqrt{25} = \sqrt{c^2}$$

$$5 = c$$

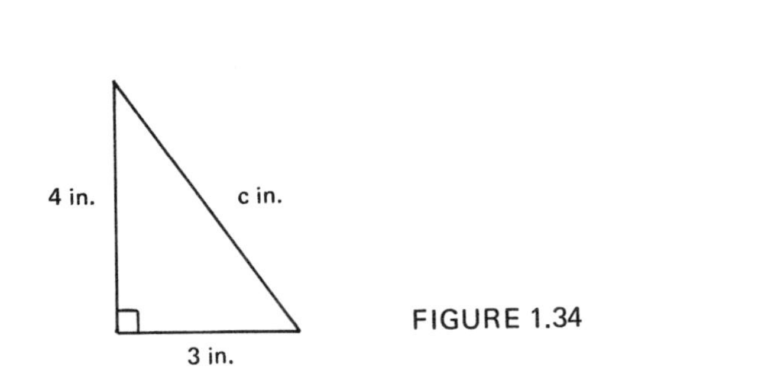

FIGURE 1.34

The hypotenuse is 5 inches.

In the following example, the hypotenuse is an irrational number.

EXAMPLE 4. Find the hypotenuse of the triangle in Figure 1.35.

SOLUTION.

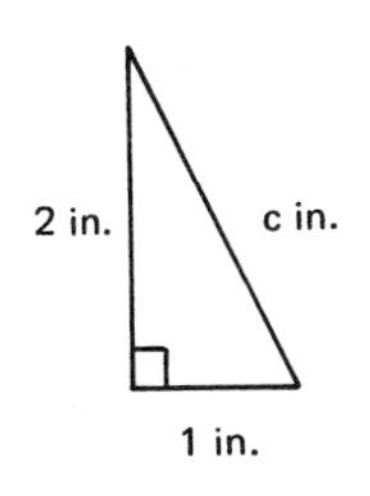

FIGURE 1.35

$$b^2 + h^2 = c^2$$

$$1^2 + 2^2 = c^2$$

$$1 + 4 = c^2$$

$$5 = c^2$$ Find the *positive* root of this equation.

$$\sqrt{5} = c$$

---

†This construction uses the fact that the sum of the measures of the angles of a triangle is 180°. (See Section 3.1.)

The *exact* measurement of the hypotenuse is $\sqrt{5}$ inches. (Usually, you will leave your result in *radical form,* as $\sqrt{5}$ inches.) Because $\sqrt{5} \approx 2.2$, the hypotenuse is *approximately* 2.2 inches.

EXAMPLE 5. Find the height of the given triangle.

SOLUTION.

$$b^2 + h^2 = c^2$$

$$6^2 + h^2 = 10^2$$

$$36 + h^2 - 36 = 100 - 36$$

$$h^2 = 64$$

$$h = 8$$

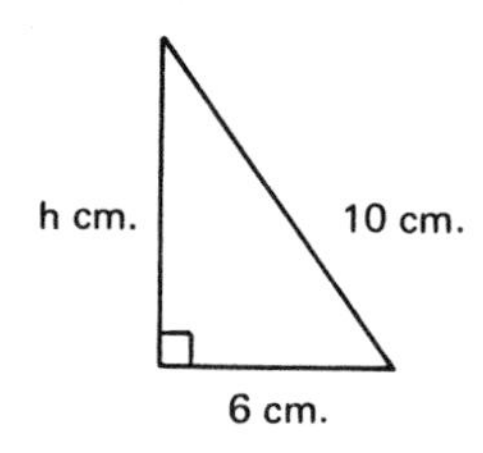


FIGURE 1.36

The height is 8 centimeters.

## CLASS EXERCISES

In Exercises 5 and 6, find the hypotenuse of the right triangle.

5. base 9 centimeters, height 12 centimeters

6. base 2 inches, height 3 inches (Leave the answer in radical form.)

7. Find the height of a right triangle with base 12 centimeters and hypotenuse 13 centimeters.

8. Find the base of a right triangle with height 5 inches and hypotenuse 6 inches. (Leave the answer in radical form.)

## PARALLEL LINES

Lines extend indefinitely.

DEFINITION. *Parallel lines* are lines (in a plane) that do not intersect.

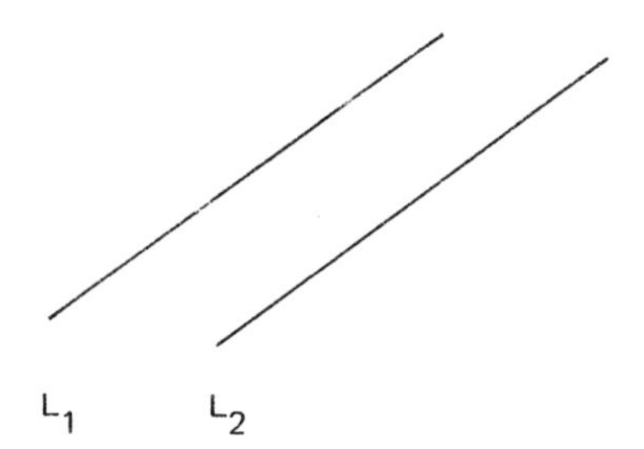

$L_1$ $L_2$

FIGURE 1.37 (a)
Lines $L_1$ and $L_2$ are parallel.

$L_3$ $L_4$

FIGURE 1.37 (b)
Lines $L_3$ and $L_4$ intersect and therefore are not parallel.

## PARALLELOGRAMS

DEFINITION. A *parallelogram* is a quadrilateral, or 4-sided polygon, in which both pairs of opposite sides are parallel. The line segment from any vertex drawn perpendicular to an "opposite side" is called a *height* of the parallelogram. The corresponding perpendicular side of the parallelogram is called a *base*. (See Figures 1.38(a) and 1.38(b).)

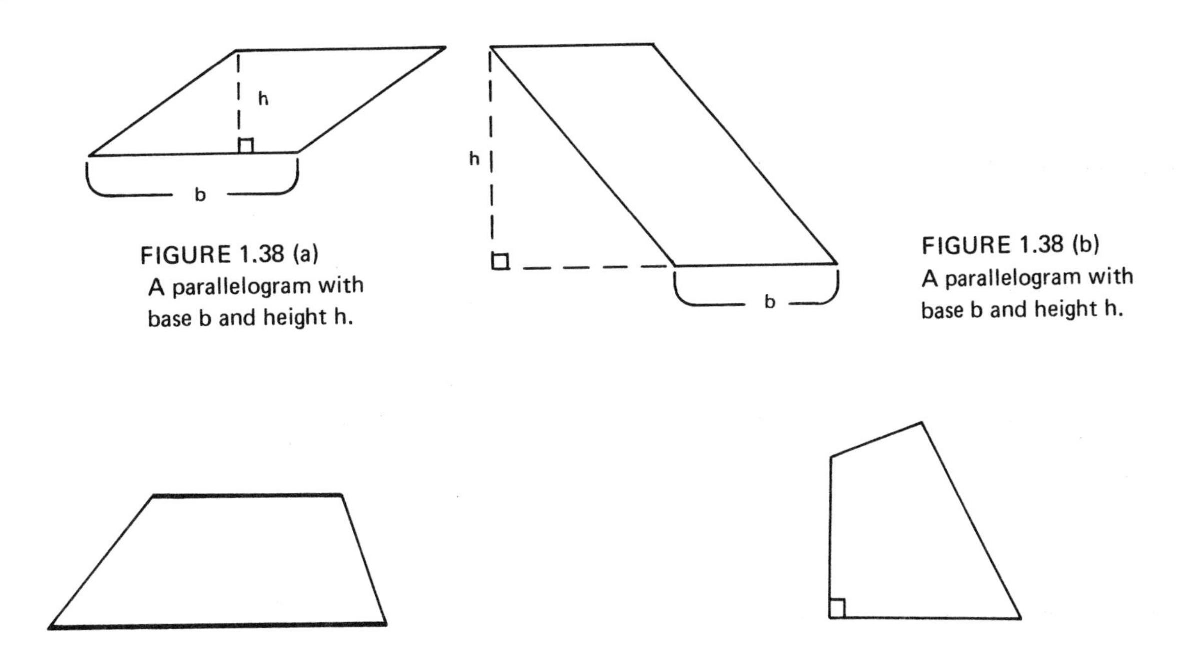


FIGURE 1.38 (a) A parallelogram with base b and height h.

FIGURE 1.38 (b) A parallelogram with base b and height h.

FIGURE 1.38 (c) This quadrilateral (which is called a trapezoid) is not a parallelogram. The sides drawn with thickened lines are parallel, but the other sides are not.

FIGURE 1.38 (d) This quadrilateral is not a parallelogram.

It can be shown that *in a parallelogram opposite sides are of equal length (and are parallel). A rectangle is a parallelogram in which adjacent sides form right angles.*

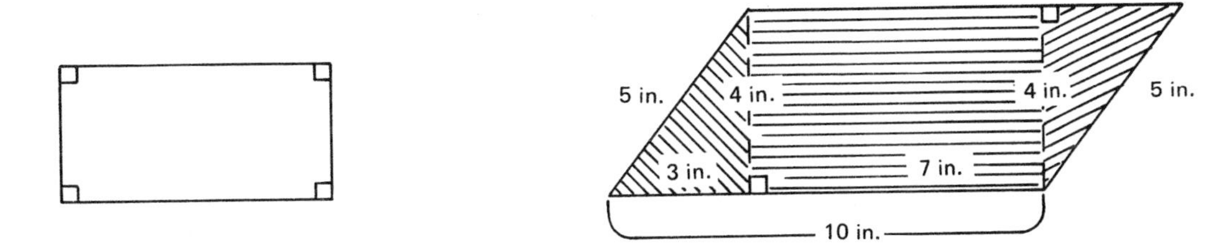


FIGURE 1.39 A rectangle is a parallelogram.

FIGURE 1.40

To find a formula for the area of a parallelogram, consider Figure 1.40. Break up the parallelogram into a rectangle and two right triangles. (We neglect units.)

Area of parallelogram = Area of rectangle + 2 × Area of each one of the triangles

$$= 7 \cdot 4 + \frac{\cancel{2}^{1}(3 \cdot 4)}{\cancel{2}_{1}}$$

$$= (7 + 3)\,4 \qquad \text{by the Distributive Laws}$$

$$= 10 \cdot 4$$

$$= 40$$

Observe that in the next to last line, 10 represents the base and 4 the height of the parallelogram.

In general,

*Area of parallelogram = base × height*

In symbols,

$$\mathbf{A} = b \cdot h$$

Note that this is the same as the formula for the area of a rectangle with base b and height h. (See Figure 1.41.)

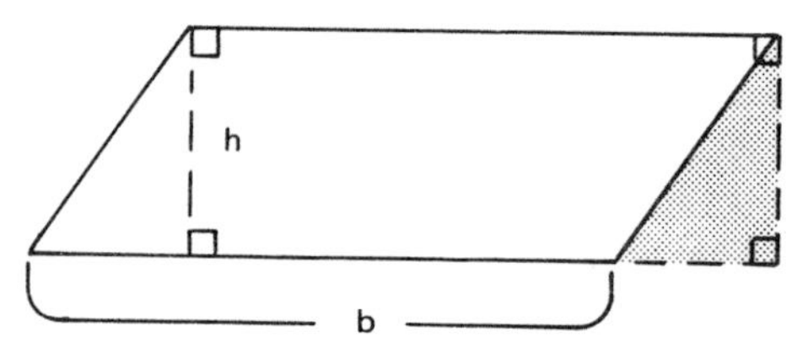


FIGURE 1.41 (a)

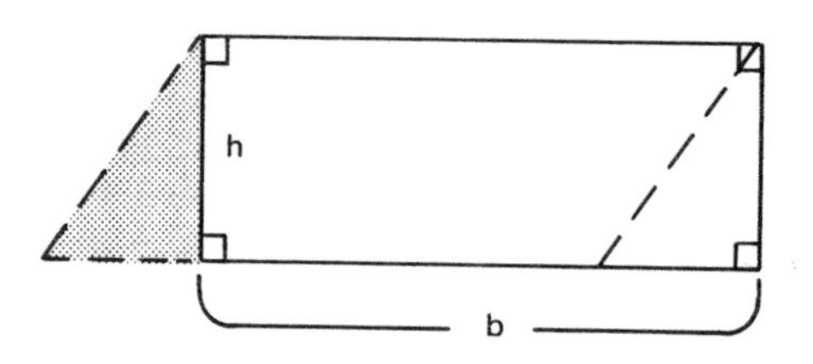


FIGURE 1.41 (b)

EXAMPLE 6. Find the area of the given parallelogram.

SOLUTION.

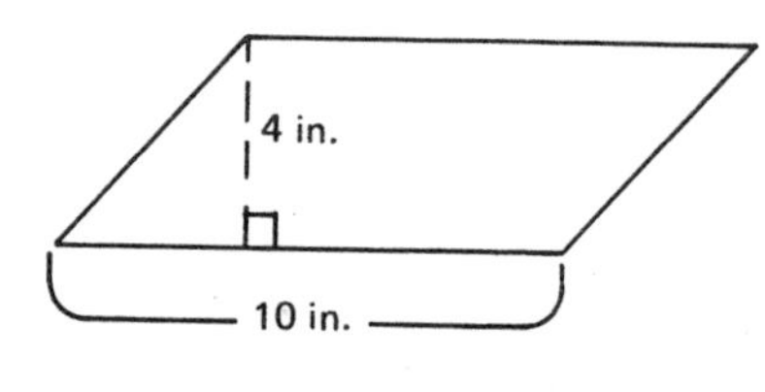


FIGURE 1.42

$A = b \cdot h$

$= 10 \text{ in.} \times 4 \text{ in.}$

$= 40 \text{ in.}^2$

## CLASS EXERCISES

Find the area of each parallelogram.

9. base 9 inches, height 7 inches

10.

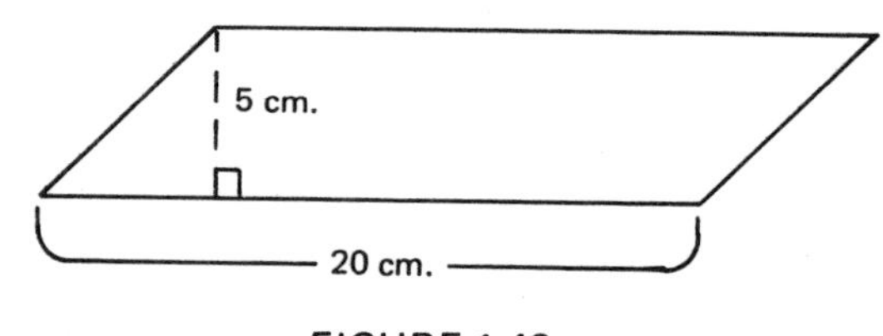


FIGURE 1.43

11.

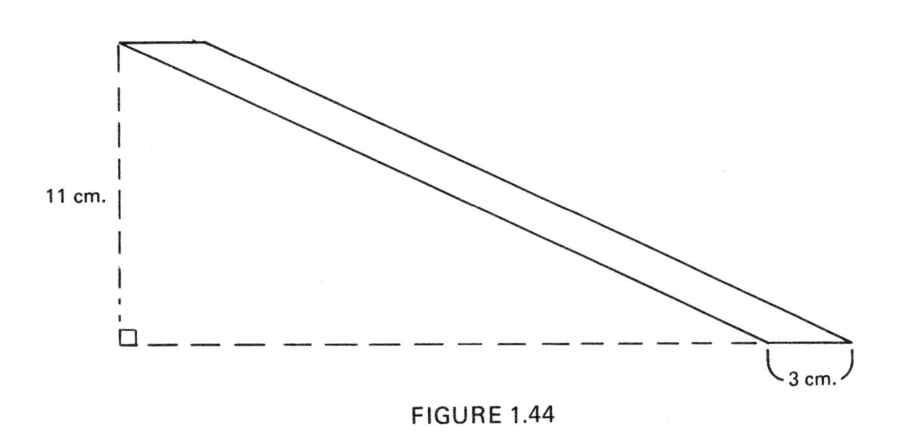


FIGURE 1.44

## TRAPEZOIDS

DEFINITION. A *trapezoid* is a quadrilateral in which exactly one pair of opposite sides is parallel.†

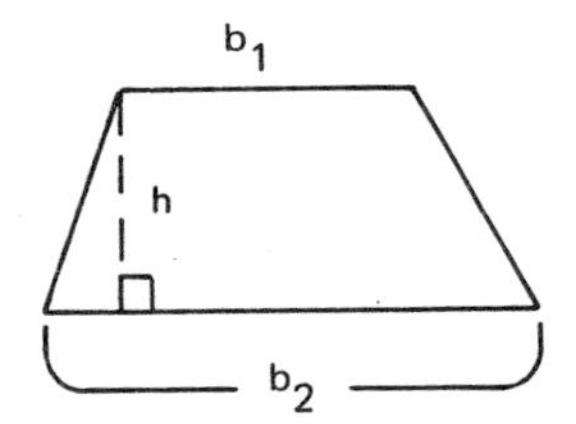


FIGURE 1.45 (a) A trapezoid with bases $b_1$ and $b_2$ and height h.

Figure 1.45 (a) depicts a trapezoid. The parallel sides are called its *bases.* The bases are unequal in length. Consider the longer base. A line segment from an "opposite vertex" drawn perpendicular to this base is called a *height* of the trapezoid. [See Figure 1.45 (a).]

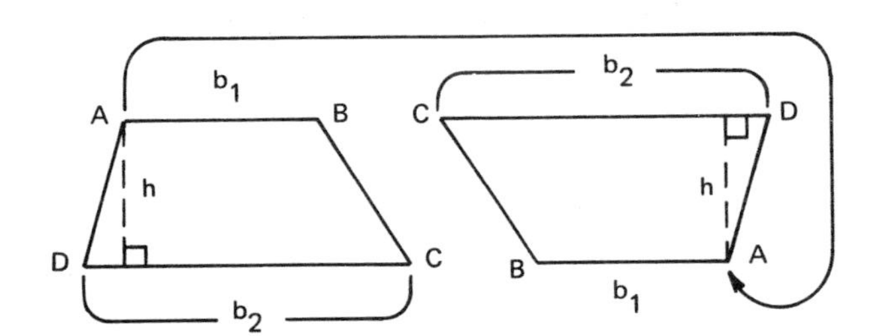


FIGURE 1.45 (b) The second trapezoid is the first one rotated. Thus, both have the same area.

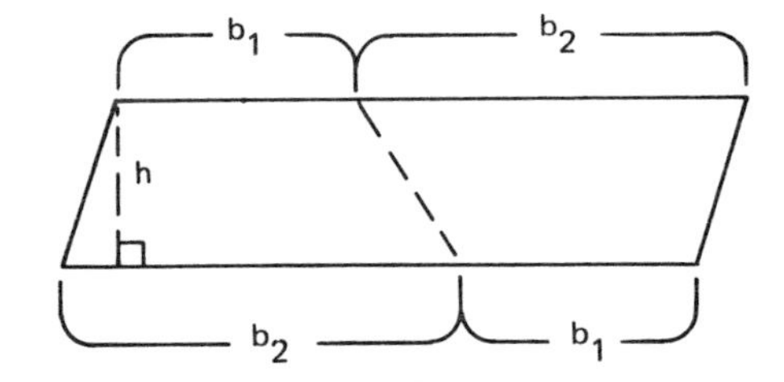


FIGURE 1.45 (c) The two trapezoids are joined to form a parallelogram with base $b_1 + b_2$ and height h.

†Note that, according to this definition, a parallelogram is *not* a trapezoid. In contrast, some authors define a trapezoid to be a quadrilateral in which *at least one* pair of opposite sides is parallel, so that a parallelogram is then a special type of trapezoid.

To find the formula for the area of a trapezoid, label the *vertices* of the trapezoid. Then rotate the trapezoid, as indicated in Figure 1.45 (b). Join the two trapezoids to form a parallelogram with base $b_1 + b_2$ and with height h, as shown in Figure 1.45 (c). The area of the parallelogram is

$$(b_1 + b_2)\,h.$$

***The area of the*** original ***trapezoid,*** which is ***half*** of that of the parallelogram, ***is***

$$\frac{(b_1 + b_2)\,h}{2}.$$

In other words, ***the area of a trapezoid is half the sum of the bases times the height.***

Note that

$$\frac{1}{2}\,(b_1 + b_2)$$

is the average of the two bases. Thus, *the area of a trapezoid is the average of the bases times the height.*

EXAMPLE 7. Find the area of the given trapezoid.

SOLUTION.

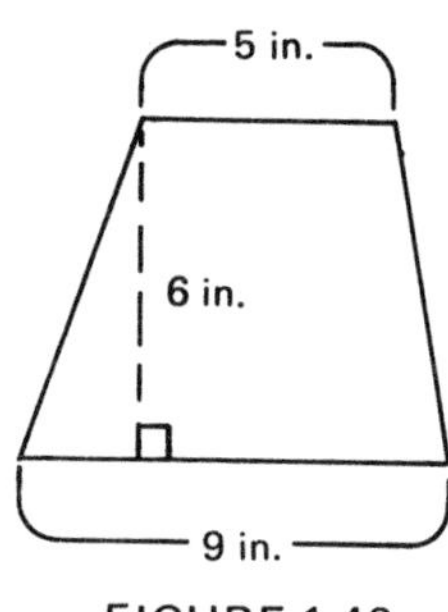


FIGURE 1.46

$$A = \frac{(b_1 + b_2)\,h}{2}$$

$$= \frac{(5 \text{ in.} + 9 \text{ in.})\ 6 \text{ in.}}{2}$$

$$= \frac{\overset{7 \text{ in.}}{\cancel{14 \text{ in.}}} \times 6 \text{ in.}}{\underset{1}{\cancel{2}}}$$

$$= 42 \text{ in.}^2$$

## CLASS EXERCISES

Find the area of each trapezoid.

12. bases 12 centimeters and 10 centimeters, height 9 centimeters

13. average of bases 8 inches, height 10 inches

14.

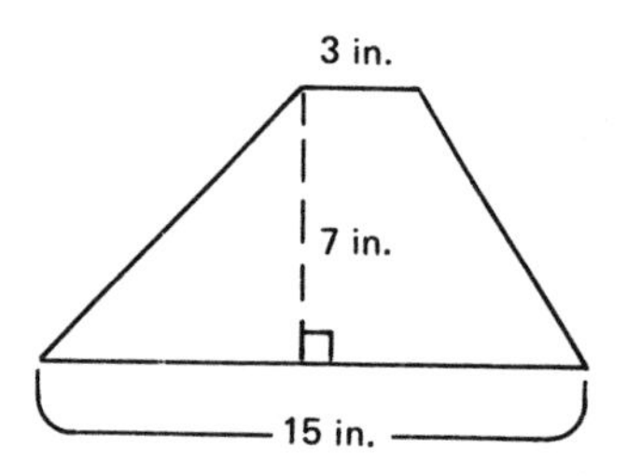


FIGURE 1.47

## SOLUTIONS TO CLASS EXERCISES

1. $A = \dfrac{b \cdot h}{2}$

$$= \frac{8 \text{ in.} \times \overset{2 \text{ in.}}{\cancel{4 \text{ in.}}}}{\underset{1}{\cancel{2}}}$$

$= 16 \text{ in.}^2$

2. $A = \dfrac{b \cdot h}{2}$

$$= \frac{\overset{2 \text{ in.}}{\cancel{4 \text{ in.}}} \times 3 \text{ in.}}{\underset{1}{\cancel{2}}}$$

$= 6 \text{ in.}^2$

3. $A = \dfrac{b \cdot h}{2}$

$$= \frac{\overset{20 \text{ cm.}}{\cancel{40 \text{ cm.}}} \times 25 \text{ cm.}}{\underset{1}{\cancel{2}}}$$

$= 500 \text{ cm.}^2$

4(a). $A = \dfrac{b \cdot h}{2}$

$$= \frac{\overset{8 \text{ cm.}}{\cancel{16 \text{ cm.}}} \times 12 \text{ cm.}}{\underset{1}{\cancel{2}}}$$

$= 96 \text{ cm.}^2$

4(b). $A = \dfrac{b \cdot h}{2}$

$$= \frac{\overset{12 \text{ cm.}}{\cancel{24 \text{ cm.}}} \times 12 \text{ cm.}}{\underset{1}{\cancel{2}}}$$

$= 144 \text{ cm.}^2$

4(c). $A = \dfrac{b \cdot h}{2}$

$$= \frac{\overset{4 \text{ cm.}}{\cancel{8 \text{ cm.}}} \times 12 \text{ cm.}}{\underset{1}{\cancel{2}}}$$

$= 48 \text{ cm.}^2$

From parts (a) and (b), note that $144 \text{ cm.}^2 - 96 \text{ cm.}^2 = 48 \text{ cm.}^2$

5. $b^2 + h^2 = c^2$

$9^2 + 12^2 = c^2$

$81 + 144 = c^2$

$225 = c^2$

$\sqrt{225} = c$

$15 = c$

The hypotenuse is 15 centimeters.

6. $b^2 + h^2 = c^2$

$2^2 + 3^2 = c^2$

$4 + 9 = c^2$

$\sqrt{13} = c$

The hypotenuse is $\sqrt{13}$ inches.

7. $b^2 + h^2 = c^2$

$12^2 + h^2 = 13^2$

$144 + h^2 = 169$

$h^2 = 25$

$h = \sqrt{25}$

$h = 5$

The height is 5 centimeters.

8. $b^2 + h^2 = c^2$

$b^2 + 5^2 = 6^2$

$b^2 + 25 = 36$

$b^2 = 11$

$b = \sqrt{11}$

The base is $\sqrt{11}$ inches.

9. $A = b \cdot h$

$= 9 \text{ in.} \times 7 \text{ in.}$

$= 63 \text{ in.}^2$

10. $A = b \cdot h$

$= 20 \text{ cm.} \times 5 \text{ cm.}$

$= 100 \text{ cm.}^2$

11. $A = b \cdot h$

$= 3 \text{ cm.} \times 11 \text{ cm.}$

$= 33 \text{ cm.}^2$

12. $A = \frac{(b_1 + b_2)h}{2}$

$= \frac{(12 \text{ cm.} + 10 \text{ cm.})\, 9 \text{ cm.}}{2}$

$= \frac{22 \text{ cm.} \times 9 \text{ cm.}}{2}$

$= 11 \text{ cm.} \times 9 \text{ cm.}$

$= 99 \text{ cm.}^2$

13. A = average of bases × height

$= 8 \text{ in.} \times 10 \text{ in.}$

$= 80 \text{ in.}^2$

14. $A = \frac{(b_1 + b_2)h}{2}$

$= \frac{(3 \text{ in.} + 15 \text{ in.})\, 7 \text{ in.}}{2}$

$= \frac{18 \text{ in.} \times 7 \text{ in.}}{2}$

$= 9 \text{ in.} \times 7 \text{ in.}$

$= 63 \text{ in.}^2$

NAME

CLASS

## HOME EXERCISES

In Exercises 1-6, find the area of each triangle.

1. base 4 inches, height 4 inches

2. base 9 inches, height 6 inches

3. base 3/4 inch, height 6 inches

4.

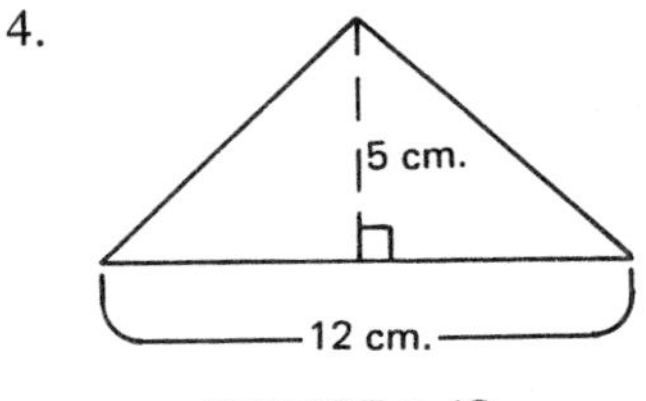


FIGURE 1.48

5.

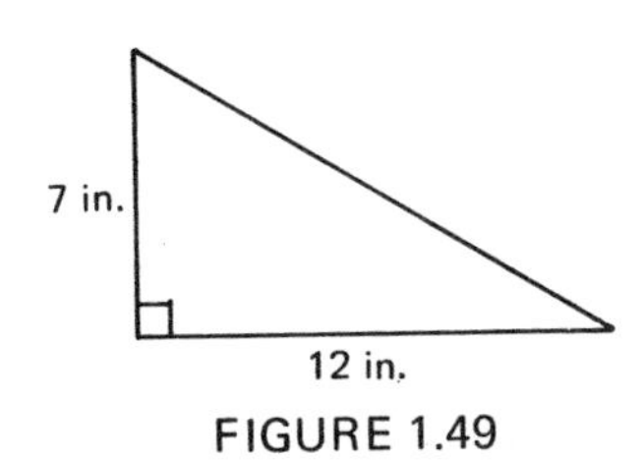


FIGURE 1.49

6.

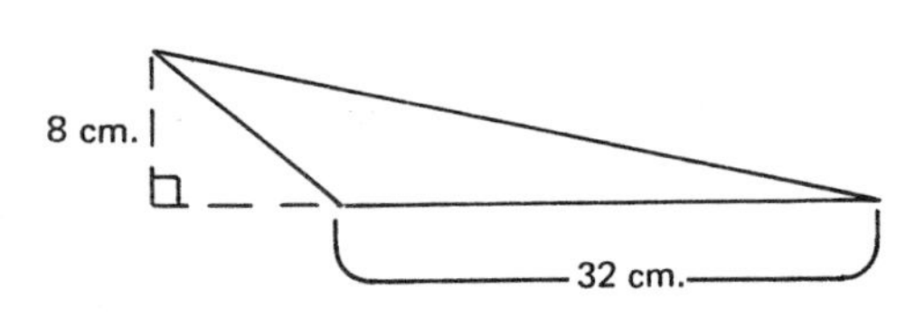


FIGURE 1.50

In Exercises 7 and 8, find the height of each triangle.

7. base 10 inches, area 100 square inches

8. base 5 centimeters, area 30 square centimeters

In Exercises 9 and 10, find the base of each triangle.

9. height 6 inches, area 9 square inches

10. height 10 feet, area 40 square feet

In Exercises 11-14, find the hypotenuse of each right triangle. If necessary, leave your answer in radical form.

11. base 40 yards, height 30 yards

12. base 8 meters, height 6 meters

ANSWERS

1. __________

2. __________

3. __________

4. __________

5. __________

6. __________

7. __________

8. __________

9. __________

10. __________

11. __________

12. __________

ANSWERS

13. ____________

14. ____________

15. ____________

16. ____________

17. ____________

18. ____________

19. ____________

20. ____________

21. ____________

22. ____________

23. ____________

24. ____________

13.

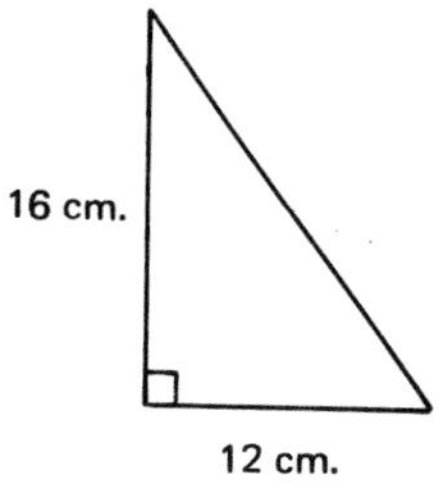

16 cm.

12 cm.

FIGURE 1.51

14.

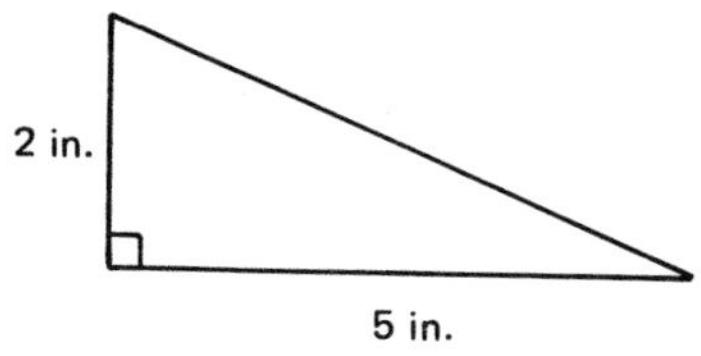

2 in.

5 in.

FIGURE 1.52

In Exercises 15-17, find the height of each right triangle. If necessary, leave your answer in radical form.

15. base 9 inches, hypotenuse 15 inches

16. base 24 inches, hypotenuse 25 inches

17. base 1 centimeter, hypotenuse 4 centimeters

In Exercises 18-20, find the base of each right triangle. If necessary, leave your answer in radical form.

18. height 16 inches, hypotenuse 20 inches

19. height 5 feet, hypotenuse 13 feet

20. height 3 feet, hypotenuse 6 feet

In Exercises 21-26, find the area of each parallelogram.

21. base 10 inches, height 8 inches

22. base 6 inches, height 6 inches

23. base 24 centimeters, height 5 centimeters

24.

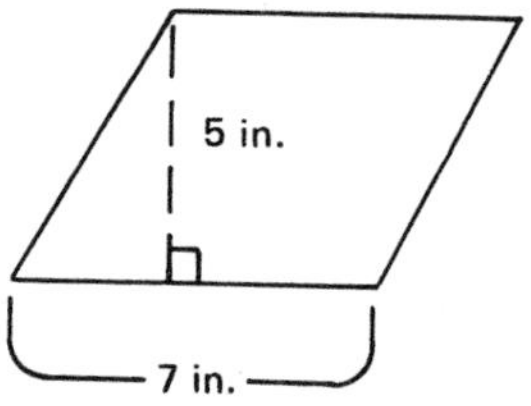

5 in.

7 in.

FIGURE 1.53 (a)

NAME

CLASS

ANSWERS

25. ____________

26. ____________

27. ____________

28. ____________

29. ____________

30. ____________

31. ____________

32. ____________

33. ____________

34. ____________

35. ____________

36. ____________

25. 26.

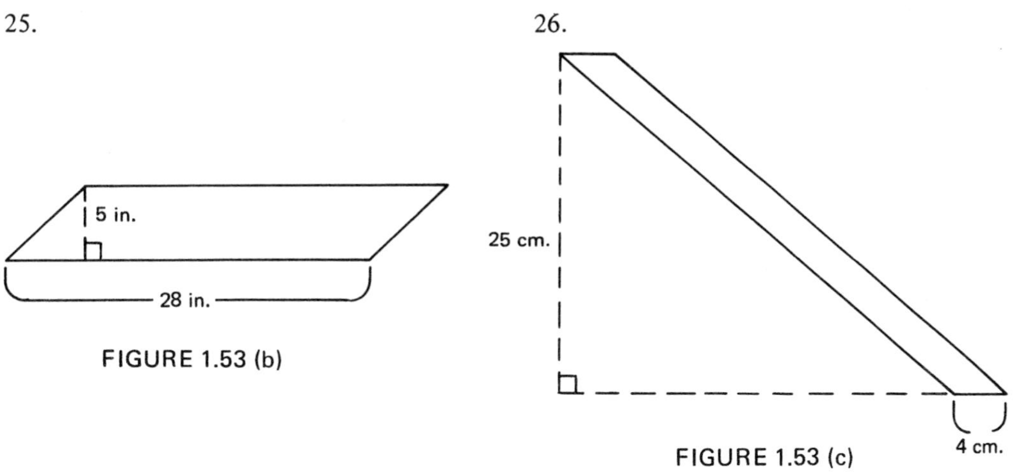


FIGURE 1.53 (b)

FIGURE 1.53 (c)

In Exercises 27 and 28, find the height of each parallelogram.

27. base 5 inches, area 45 square inches

28. base 8 decimeters, area 32 square decimeters

In Exercises 29 and 30, find the base of each parallelogram.

29. height 9 inches, area 72 square inches

30. height 15 centimeters, area 75 square centimeters

In Exercises 31-36, find the area of each trapezoid.

31. bases 4 inches and 6 inches, height 5 inches

32. average of bases 4 meters, height 1 meter

33. bases 1/2 inch and 3/2 inches, height 3 inches

34. bases 10 inches and 1 foot, height 10 inches

35.

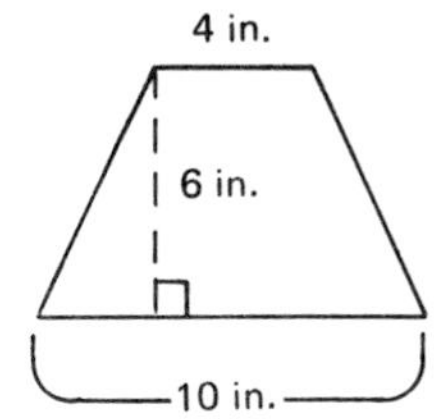


FIGURE 1.54

36.

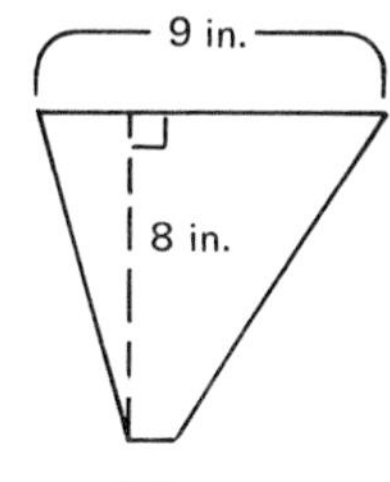


FIGURE 1.55

ANSWERS

37. ____________________

38. ____________________

39. ____________________

40. ____________________

41. ____________________

42. ____________________

43. ____________________

44. ____________________

In Exercises 37-39, find the height of each trapezoid.

37. bases 6 inches and 8 inches, area 77 square inches

38. bases 10 centimeters and 20 centimeters, area 30 square centimeters

39. average of bases 9 inches, area 70 square inches

40. The height of a trapezoid is 5 inches and its area is 40 square inches. Find the sum of the bases.

41. The height of a trapezoid is 7 inches and its area is 28 square inches. Find the average of the bases.

42. The height of a trapezoid is 12 centimeters, the area is 60 square centimeters, and one base is 6 centimeters. Find the other base.

In Exercises 43-47, find the area of each figure. (These figures consist of rectangles, triangles, parallelograms, and trapezoids.)

43.

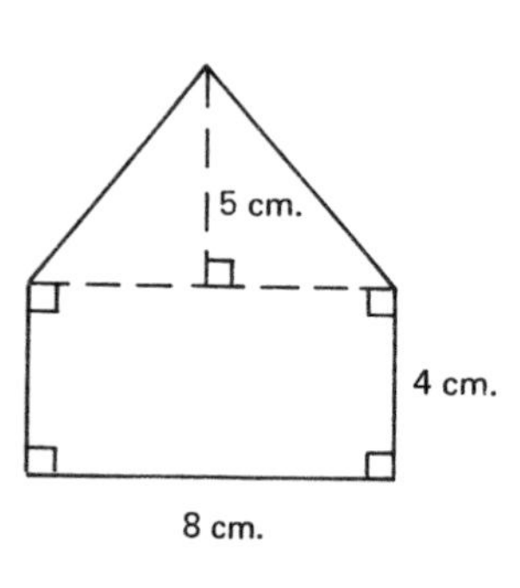


FIGURE 1.56

44.

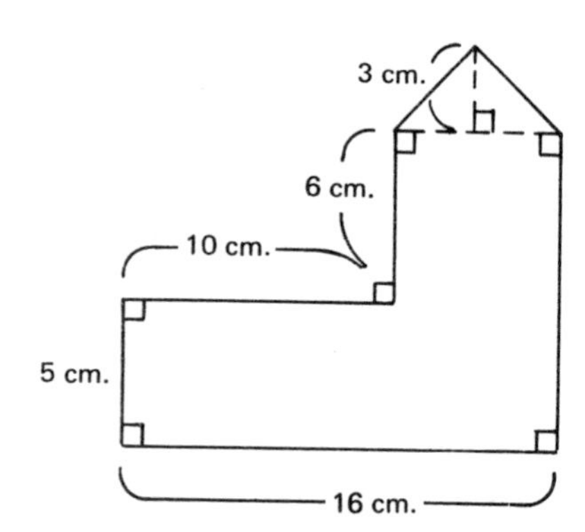


FIGURE 1.57

NAME

CLASS

ANSWERS

45. ____________

46. ____________

47. ____________

45.

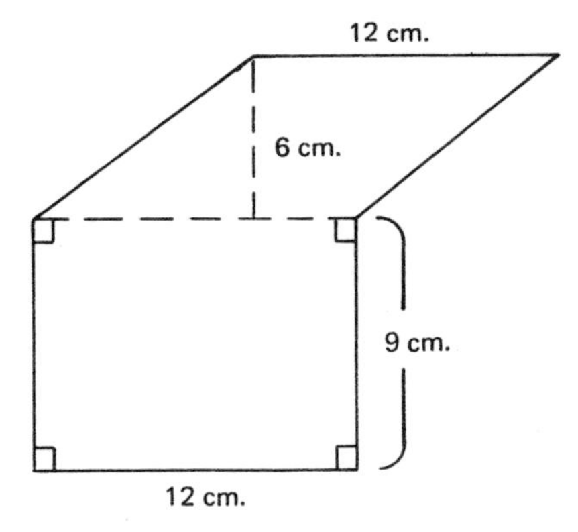


FIGURE 1.58 (a)

46.

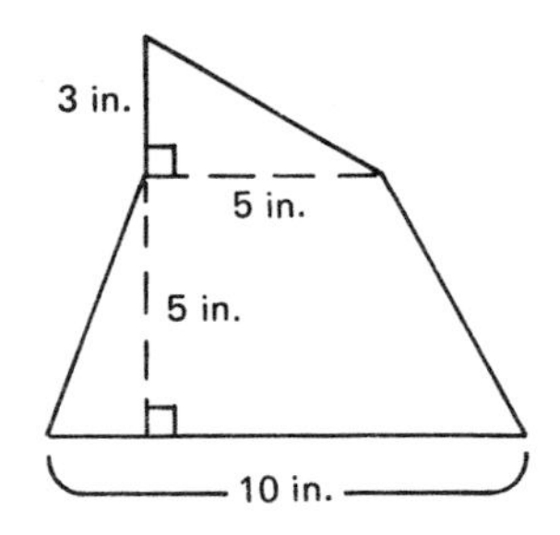


FIGURE 1.58 (b)

47.

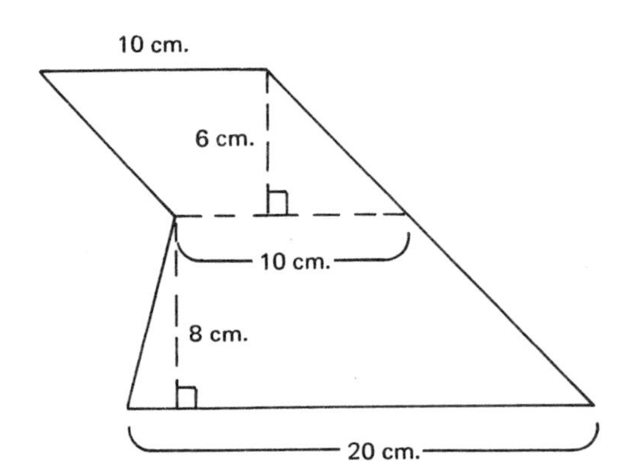


FIGURE 1.58 (c)

## 1.4 Circles

### RADIUS AND DIAMETER

DEFINITION. A *circle* is closed figure every point of which is at a fixed distance from a given point, known as the *center* of the circle. The fixed distance is called the *radius* of the circle. (The plural of "radius" is "radii.")

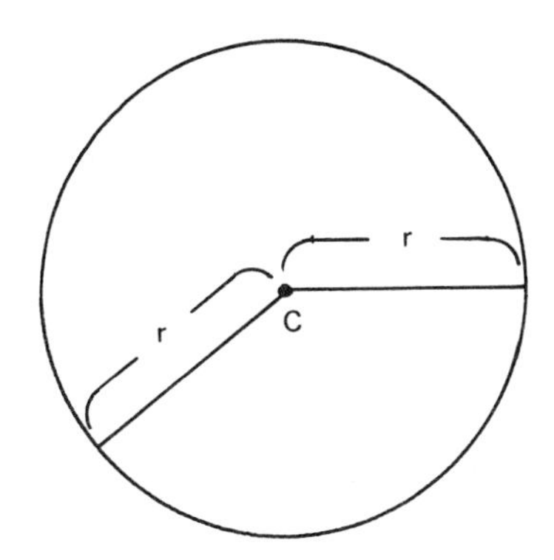


FIGURE 1.59 A circle with center C and radius r.

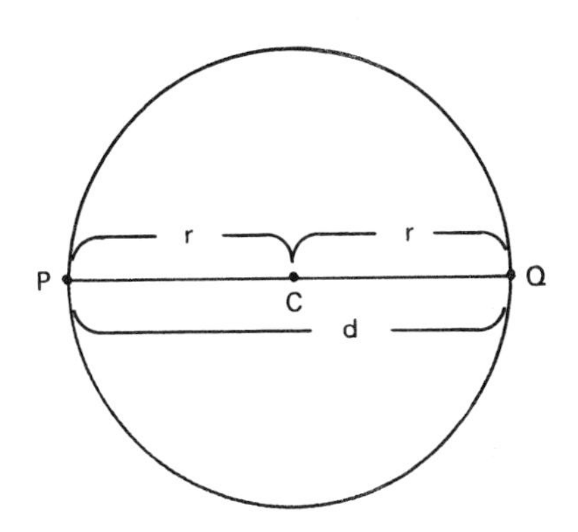


FIGURE 1.60 A circle with diameter d. d = 2r

Consider any line through the center of a circle. It cuts the circle at two points, P and Q. (See Figure 1.60.) The length of the line segment from P to Q is called the *diameter* of the circle. Clearly, the *diameter is twice the radius.* In symbols, let

$$d = \text{diameter},$$
$$r = \text{radius}.$$

Then
$$d = 2r$$

Thus, if the radius of a circle is 7 inches, its diameter is 14 inches.

EXAMPLE 1. Find the radius of the given circle.

SOLUTION.

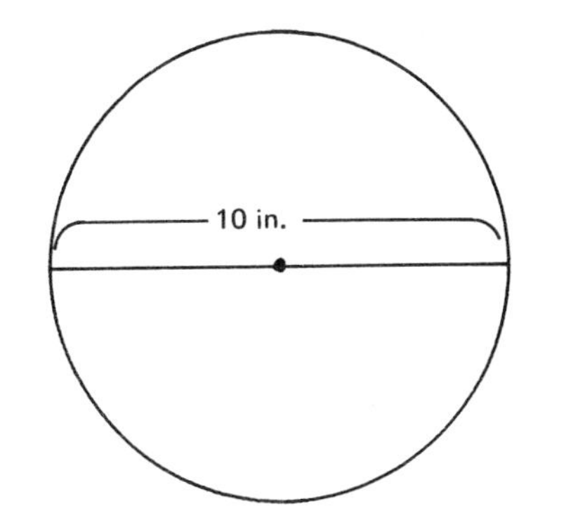


FIGURE 1.61

$$d = 2\,r$$
$$10 \text{ in.} = 2\,r$$
$$\frac{10 \text{ in.}}{2} = \frac{2\,r}{2}$$
$$5 \text{ in.} = r$$

## CLASS EXERCISES

1. Find the diameter of the circle with radius 6 inches.

2. Find the radius of the circle with diameter 6 inches.

### CIRCUMFERENCE

Recall that the perimeter of a polygon can be thought of as the distance around it. The *circumference* of a circle is defined to be the distance around the circle. The ancient Greeks discovered that *the circumference of any circle is always equal to the diameter times the irrational number* $\pi$. To the nearest hundredth,

$$\pi \approx 3.14$$

Let

$$C = \text{circumference.}$$

Then

$$C = \pi\, d$$

For a circle of diameter 5 inches, the circumference, C, is $5\pi$ inches.

$$C \approx \underbrace{5 \times 3.14}_{15.70} \text{ in.}$$

To the nearest tenth of an inch,

$$C \approx 15.7 \text{ inches}$$

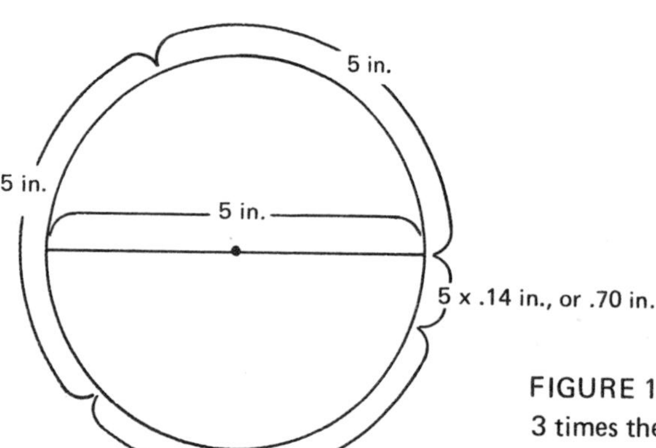


FIGURE 1.62 The circumference is (approximately) 3 times the diameter plus .14 times the diameter.

Because

$$C = \pi d$$

and

$$d = 2r,$$

replace d by 2r in the first formula to obtain

$$C = \pi \cdot 2r$$

or

$$C = 2\pi r.$$

Thus, *the circumference of a circle is $2\pi$ times the radius.*

*When approximations involving $\pi$ are called for, if nothing to the contrary is stated, use*

$$\pi \approx 3.14$$

EXAMPLE 2. Find the circumference of a circle of radius 10 centimeters.

(a) Express your answer in terms of $\pi$.

(b) Approximate to the nearest centimeter.

SOLUTION. (a) $C = 2\pi r$

$= 2\pi \times 10$ cm.

$= 20\pi$ cm.

(b) $C \approx \underbrace{20 \times 3.14}_{62.80}$ cm.

To the nearest centimeter,
$C \approx 63$ cm.

## CLASS EXERCISES

Find the circumference of the circle. (a) Express your answer in terms of $\pi$. (b) Approximate to the nearest tenth of a unit.

3. diameter 4 inches

4. radius 5 centimeters

### AREA

Consider a circle of radius r. Draw the square that has side length r, as shown in Figure 1.63 (a). About 3/4 of the square region lies within the circle. Because the area of the square is $r^2$, the area of this pie-shaped circular region in Figure 1.63 (a) is *approximately* $\frac{3}{4} r^2$.

Now, draw 4 of these squares, as in Figure 1.63 (b). The total area of the 4-square region is $4r^2$. The area of the circular region is therefore *approximately*

$$\frac{3}{4} \cdot 4r^2, \text{ or } 3r^2.$$

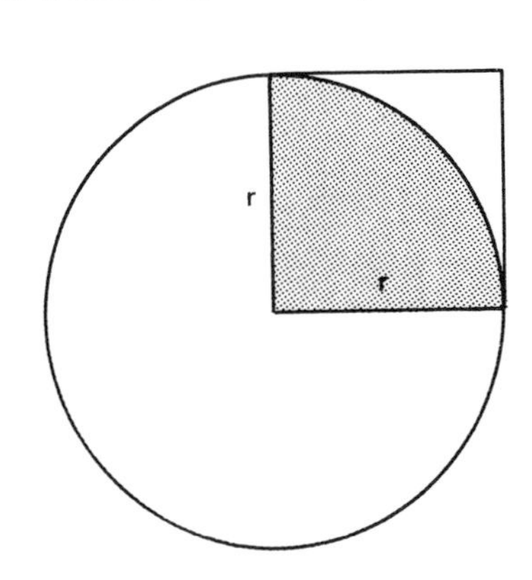

FIGURE 1.63 (a)

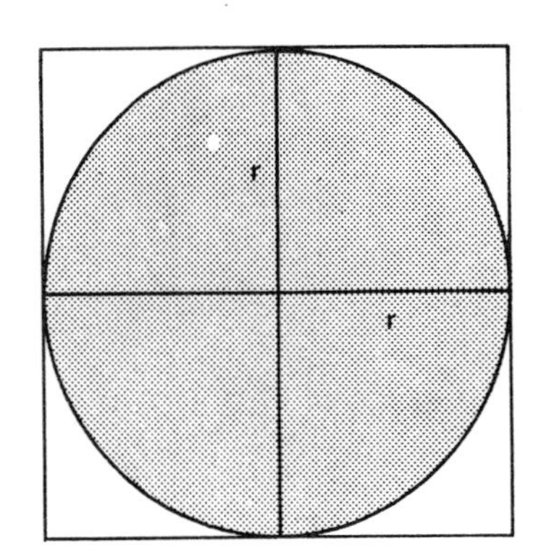

FIGURE 1.63 (b)

The Greeks discovered that **the *exact* area, A, of a circle is given by the formula**

$$A = \pi r^2.$$

Thus,

$$A \approx 3.14r^2$$

EXAMPLE 3. Find the area of a circle with radius 10 centimeters.

(a) Express your answer in terms of $\pi$.

(b) Approximate to the nearest centimeter, using $\pi \approx 3.14$.

SOLUTION. (a)

$$A = \pi r^2$$
$$= \pi(10 \text{ cm.})^2$$
$$= \pi \times 100 \text{ cm.}^2$$
$$= 100\,\pi \text{ cm.}^2$$

(b)

$$A \approx 100 \times 3.14 \text{ cm.}^2$$
$$= 314 \text{ cm.}^2$$

## CLASS EXERCISES

Find the area of each circle.

(a) Express your answer in terms of $\pi$.

(b) Approximate to the nearest unit.

5. radius = 7 in.

6. diameter = 18 cm.

## SOLUTIONS TO CLASS EXERCISES

1. $d = 2r$

$$= 2 \times 6 \text{ in.}$$
$$= 12 \text{ in.}$$

2. $d = 2r$

$$6 \text{ in.} = 2r$$
$$\frac{6 \text{ in.}}{2} = \frac{2r}{2}$$
$$3 \text{ in.} = r$$

3.(a) $C = \pi d$

$$= \pi \times 4 \text{ in.}$$
$$= 4\pi \text{ in.}$$

4.(a) $C = 2\pi r$

$$= 2\pi \times 5 \text{ cm.}$$
$$= 10\pi \text{ cm.}$$

3.(b) $C \approx \underbrace{4 \times 3.14}_{12.56}$ in.

To the nearest tenth of an inch,

$C \approx 12.6$ in.

4.(b) $C \approx 10 \times 3.14$ cm.

$= 31.4$ cm.

5.(a) $A = \pi r^2$

$= \pi \times (7 \text{ in.})^2$

$= 49\pi \text{ in.}^2$

(b) $A \approx \underbrace{49 \times 3.14}_{153.86} \text{ in.}^2$

To the nearest square inch,

$A \approx 154 \text{ in.}^2$

6.(a) $d = 2r$

$18 \text{ cm.} = 2r$

$9 \text{ cm.} = r$

$A = \pi r^2$

$= \pi(9 \text{ cm.})^2$

$= 81\pi \text{ cm.}^2$

(b) $A \approx \underbrace{81 \times 3.14}_{254.34} \text{ cm.}^2$

To the nearest square centimeter,

$A \approx 254 \text{ cm.}^2$

NAME

CLASS

## HOME EXERCISES

In approximations, unless otherwise specified, use $\pi \approx 3.14$.
In Exercises 1-4, find the diameter of the given circle.

1. radius 10 inches

2. radius 5 feet

3.

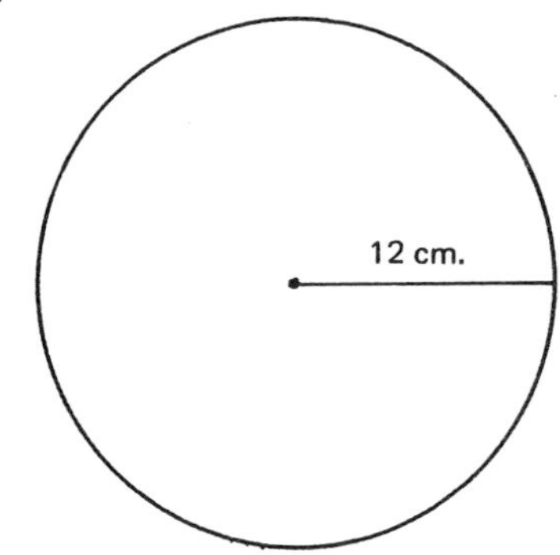


FIGURE 1.64

4.

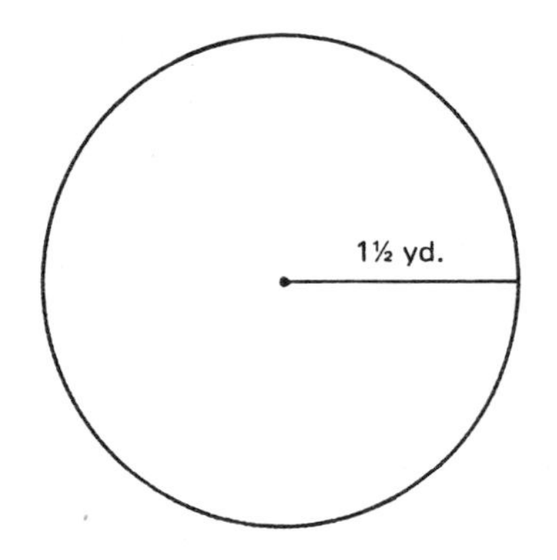


FIGURE 1.65

In Exercises 5-8, find the radius of the given circle.

5. diameter 4 inches

6. diameter 2 miles

7.

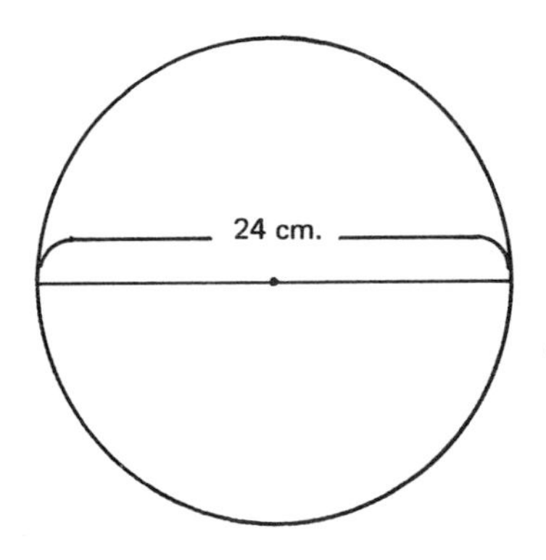


FIGURE 1.66

8.

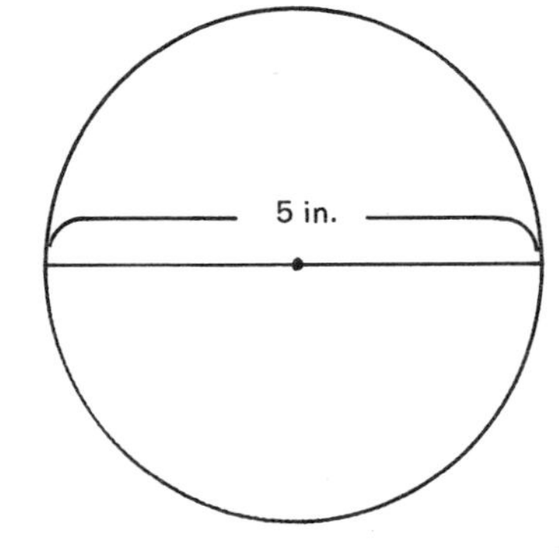


FIGURE 1.67

In Exercises 9-18, find the circumference of the given circle.

(a) Express your answer in terms of $\pi$.

(b) Approximate to the nearest unit.

9. diameter 1 inch

10. diameter 12 centimeters

11. diameter 9 yards

ANSWERS

1. ____

2. ____

3. ____

4. ____

5. ____

6. ____

7. ____

8. ____

9. (a) ____

(b) ____

10. (a) ____

(b) ____

11. (a) ____

(b) ____

ANSWERS

12. (a) ____________
 (b) ____________
13. (a) ____________
 (b) ____________
14. (a) ____________
 (b) ____________
15. (a) ____________
 (b) ____________
16. (a) ____________
 (b) ____________
17. (a) ____________
 (b) ____________
18. (a) ____________
 (b) ____________
19. (a) ____________
 (b) ____________
20. (a) ____________
 (b) ____________
21. (a) ____________
 (b) ____________
22. (a) ____________
 (b) ____________
23. (a) ____________
 (b) ____________
24. (a) ____________
 (b) ____________
25. (a) ____________
 (b) ____________
26. (a) ____________
 (b) ____________
27. (a) ____________
 (b) ____________
28. (a) ____________
 (b) ____________

12. radius 50 inches 13. radius 1 foot 14. radius 20 meters 15. radius 1/2 mile

16. 17. 18.

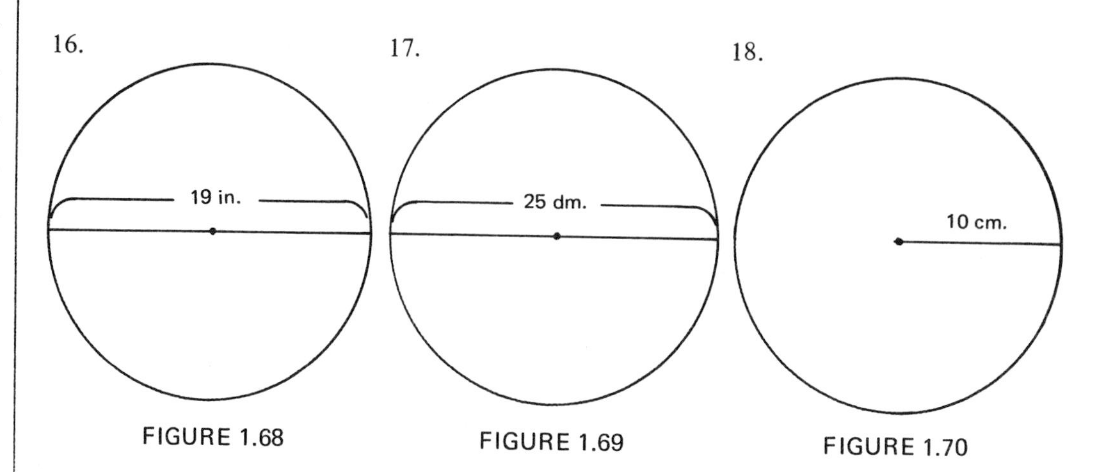


FIGURE 1.68 FIGURE 1.69 FIGURE 1.70

In Exercises 19-28, find the area of each circle.

(a) Express your answer in terms of $\pi$.

(b) Approximate to the nearest unit.

19. radius 3 yards

20. radius 5 feet

21. radius 20 inches
[In part (b), use $\pi \approx 3.142$.]

22. radius 50 centimeters
[In part (b), use $\pi \approx 3.1416$.]

23. diameter 12 dekameters

24. diameter 40 feet
[In part (b), use $\pi \approx 3.142$.]

25. diameter 200 miles
[In part (b), use $\pi \approx 3.1416$.]

26. [In part (b), use $\pi \approx 3.142$.]

27. [In part (b), use $\pi \approx 3.142$.]

28. [In part (b), use $\pi \approx 3.1416$.]

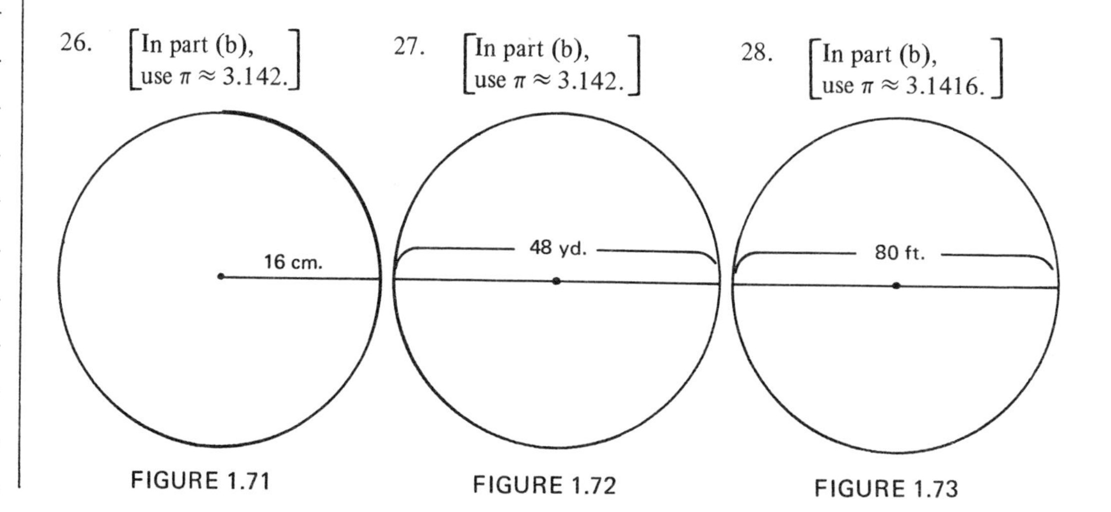


FIGURE 1.71 FIGURE 1.72 FIGURE 1.73

NAME ____________________

CLASS ____________________

ANSWERS

29. Find the area of a circle whose circumference is $6\pi$ inches.

30. Find the circumference of a circle whose area is $25\pi$ square feet.

31. Suppose the circumference of a circle is $10\pi$ centimeters. Find (a) the diameter, (b) the radius, (c) the area.

32. Suppose the area of a circle is $9\pi$ square feet. Find (a) the radius, (b) the diameter, (c) the circumference.

33. A circular swimming pool is 60 feet in diameter. How far is it around the pool? Approximate to the nearest foot.

34. A small (circular) pizza pie has a radius of 6 inches and costs one dollar. A large pizza pie has a radius of 9 inches and costs two dollars. Which is the better buy?

35. Find the area of each shaded region in Figure 1.74. Leave your answers in terms of $\pi$.

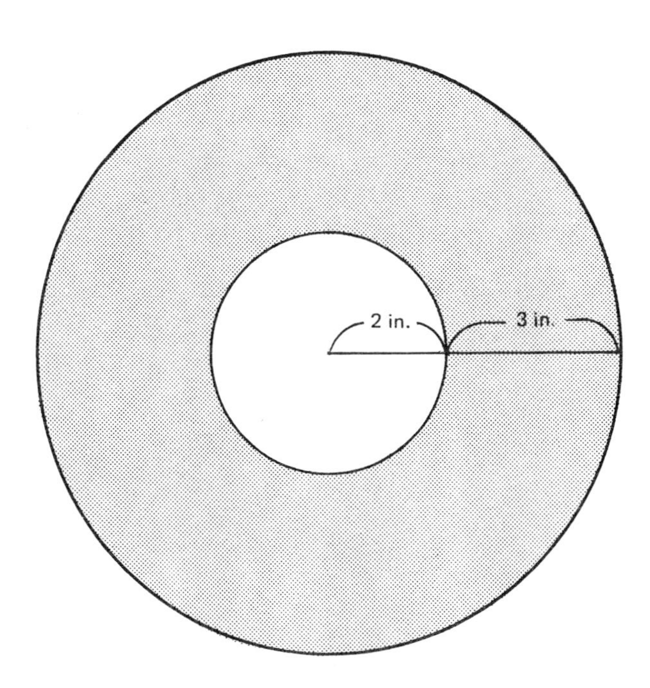


FIGURE 1.74 (a)

29. ____________________

30. ____________________

31. (a) ____________________

(b) ____________________

(c) ____________________

32. (a) ____________________

(b) ____________________

(c) ____________________

33. ____________________

34. ____________________

35. (a) ____________________

ANSWERS

(b) ______________

(c) ______________

(d) ______________

FIGURE 1.74 (b)

FIGURE 1.74 (c)

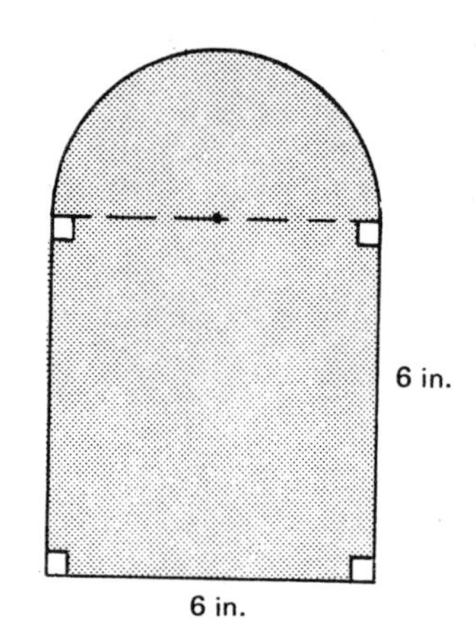


FIGURE 1.74 (d)

36. ______________

36. The diameter of a sink faucet is 1/2 inch, whereas the diameter of a bathtub faucet is 1 inch. The water pressure is the same in both cases. How much faster does the water flow from the bathtub faucet than from the sink faucet?

## 1.5 Similar Figures

### SHAPE

Intuitively, two geometric figures are said to be *similar* if they have the same shape. (Their sizes may possibly be different.) If the figures have the same size as well as the same shape, they are said to be *congruent.* We will be concerned with similar figures that are not congruent. The rectangles in Figure 1.75 are similar; those in Figure 1.76 are not. The triangles in Figure 1.77 are similar; those in Figure 1.78 are not.

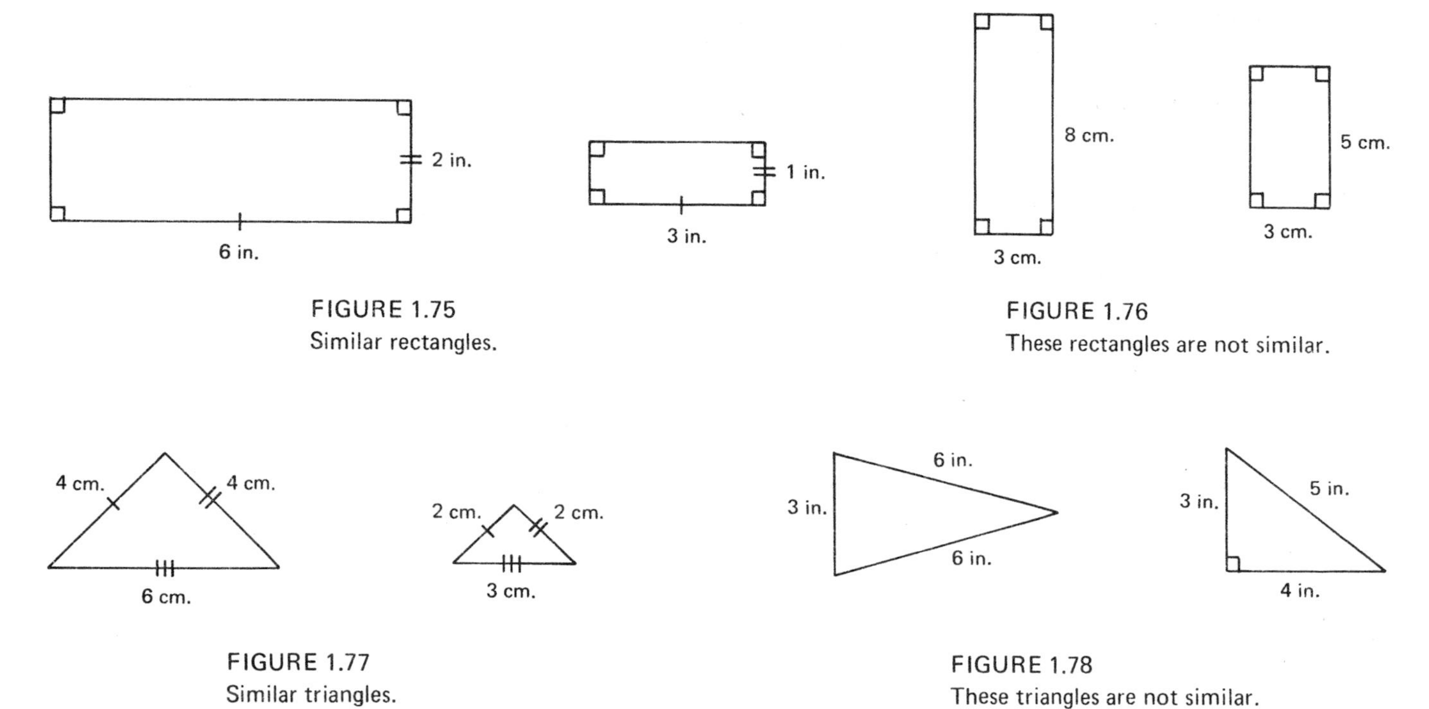


FIGURE 1.75
Similar rectangles.

FIGURE 1.76
These rectangles are not similar.

FIGURE 1.77
Similar triangles.

FIGURE 1.78
These triangles are not similar.

### PROPORTIONS

Recall that a *proportion is a statement that two fractions are equal.* Thus,

$$\frac{1}{3} = \frac{2}{6} \text{ and } \frac{5}{20} = \frac{c}{8}$$

are each proportions.

You can solve the proportion

$$\frac{5}{20} = \frac{c}{8}$$

for c by cross-multiplying:

$$5 \cdot 8 = 20c$$

$$\frac{40}{20} = \frac{20c}{20}$$

$$2 = c$$

## RECTANGLES

Two rectangles are similar if the lengths of their corresponding sides are proportional. In Figure 1.75, corresponding sides are marked with the same number of bars.

Let one rectangle have

length $\ell$, width $w$;

and let the other rectangle have

length $\ell'$, width $w'$.

Then the *two rectangles are similar if*

$$\frac{\ell}{\ell'} = \frac{w}{w'}.$$

EXAMPLE 1. Show that the given rectangles are similar.

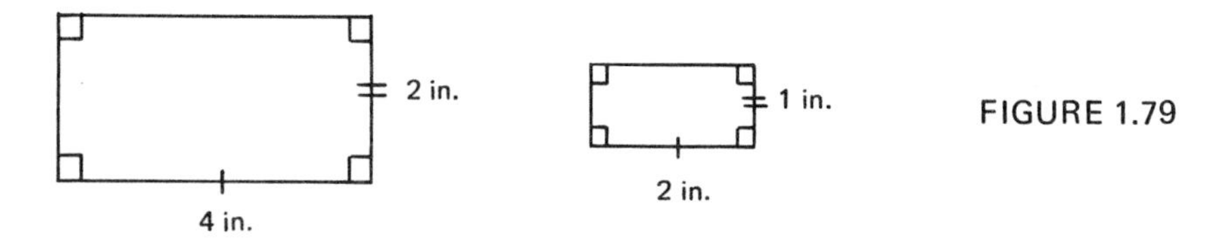


FIGURE 1.79

SOLUTION. You must show that $\dfrac{\ell}{\ell'} = \dfrac{w}{w'}$.

$$\frac{4 \text{ in.}}{2 \text{ in.}} \stackrel{?}{=} \frac{2 \text{ in.}}{1 \text{ in.}}$$

Instead of cross-multiplying, it is easiest to divide the numerator and denominator of each fraction by common factors and by common units.

$$\frac{\overset{2}{\cancel{4}}\,\cancel{\text{in.}}}{\underset{1}{\cancel{2}}\,\cancel{\text{in.}}} \stackrel{?}{=} \frac{2\,\cancel{\text{in.}}}{1\,\cancel{\text{in.}}}$$

$$2 \stackrel{\checkmark}{=} 2$$

When two rectangles are similar, the ratio $\frac{\ell}{\ell'}$, is called the *ratio of similarity*. In Example 1, the ratio of similarity is 2. *We will always take* $\ell > \ell'$, *so that the ratio of similarity will be greater than* 1. (For congruent rectangles, $\ell = \ell'$ and the ratio of similarity is 1.)

EXAMPLE 2. Suppose the given rectangles are similar.

(a) Find $w$. (b) What is the ratio of similarity?

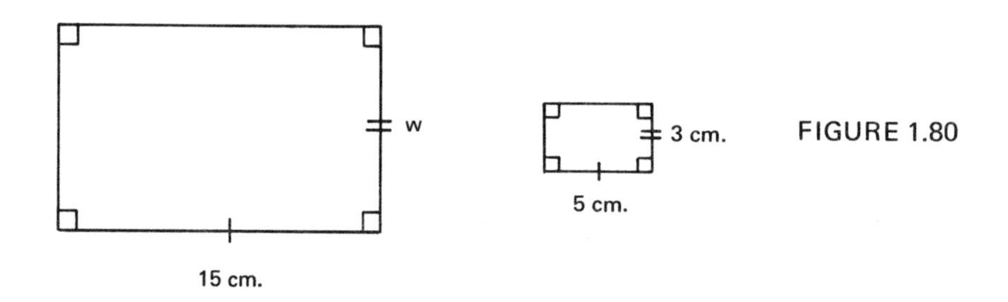


FIGURE 1.80

SOLUTION.

(a)

$$\frac{\overset{3}{\cancel{15}}\ \cancel{\text{cm.}}}{\underset{1}{\cancel{5}}\ \cancel{\text{cm.}}} = \frac{w}{3\text{ cm.}}$$

$$3 \times 3\text{ cm.} = w$$

$$9\text{ cm.} = w$$

(b) The ratio of similarity is given by

$$\frac{\ell}{\ell'} = \frac{\overset{3}{\cancel{15}}\ \cancel{\text{cm.}}}{\underset{1}{\cancel{5}}\ \cancel{\text{cm.}}}$$

$$= 3.$$

Suppose two rectangles are similar. How do their areas compare? In Example 2, the area of the first rectangle is given by

$$A = 15\text{ cm.} \times 9\text{ cm.}$$

$$= 135\text{ cm.}^2$$

and the area of the second rectangle by

$$A' = 5\text{ cm.} \times 3\text{ cm.}$$

$$= 15\text{ cm.}^2$$

Observe that

$$\frac{A}{A'} = \frac{\overset{9}{\cancel{135}}\ \cancel{\text{cm.}^2}}{\underset{1}{\cancel{15}}\ \cancel{\text{cm.}^2}}$$

$$= 9$$

$$= 3^2.$$

Thus, the ratio of similarity is 3 and the ratio of the area is $3^2$.

In general, *if the ratio of similarity of two rectangles is* R, *then the ratio of their areas is* $R^2$.

## CLASS EXERCISES

1. (a) Show that the given rectangles are similar.

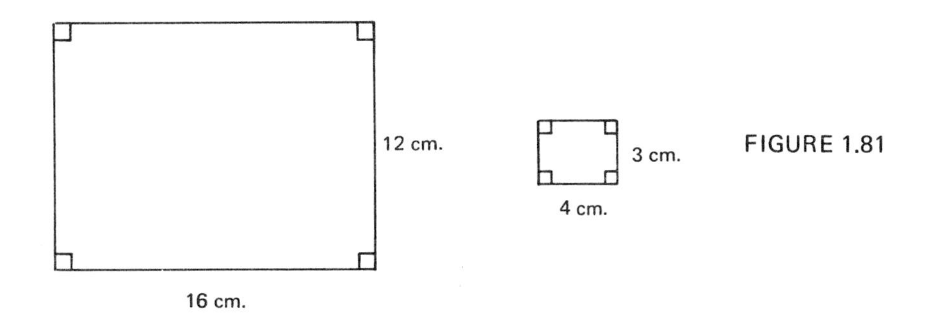


FIGURE 1.81

(b) What is the ratio of similarity?

2. Show that the given rectangles are *not* similar.

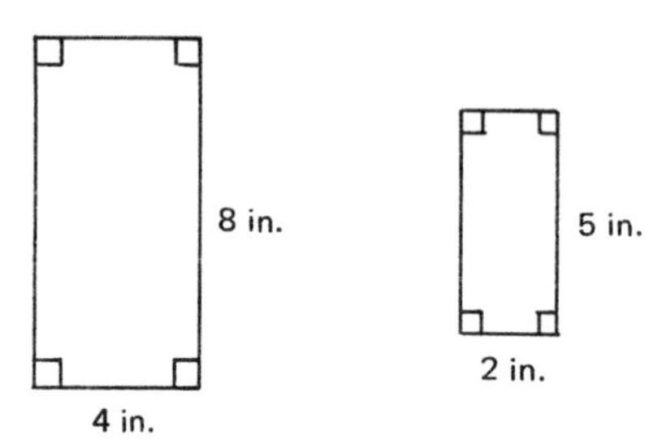


FIGURE 1.82

3. Suppose the given rectangles are similar. (a) Find $w$. (b) What is the ratio of similarity? (c) What is the ratio of the areas?

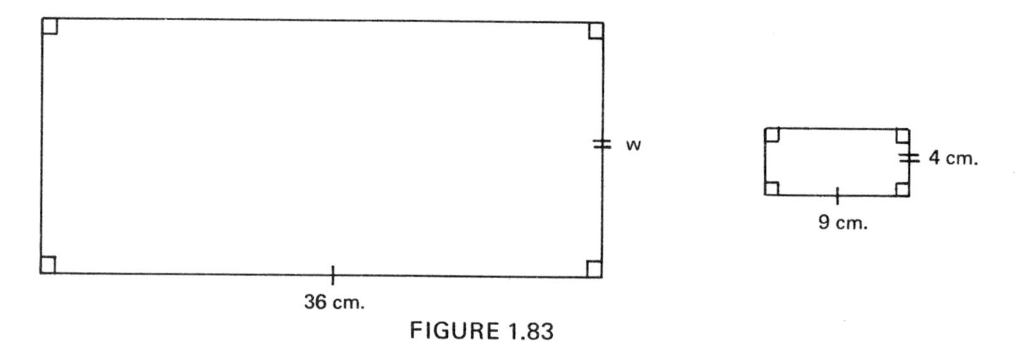


FIGURE 1.83

## SQUARES AND CIRCLES

Any two squares have the same shape, and are therefore similar. Suppose the side length of one square is $\ell$ and of the other square is $\ell'$, where $\ell > \ell'$. Because $\ell = w$ (and $\ell' = w'$), it follows that

$$\frac{\ell}{\ell'} = \frac{w}{w'} .$$

*The ratio of similarity is given by* $\frac{\ell}{\ell'}$*, and the ratio of the areas by* $\left(\frac{\ell}{\ell'}\right)^2$.

In Figure 1.84, the ratio of similarity as $\frac{3}{2}$ and the ratio of the areas is $\frac{9}{4}$.

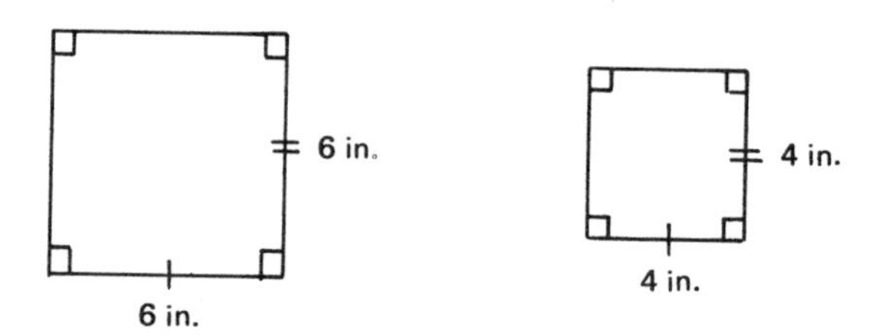


FIGURE 1.84

Any two circles have the same shape, and are thus similar. Let the radii of the circles be r and r$'$, where r > r$'$. Then *the ratio of similarity is given by* $\frac{r}{r'}$ *and the ratio of the areas is given by* $\left(\frac{r}{r'}\right)^2$.

EXAMPLE 3. (a) Find the ratio of similarity of the given circles.

(b) Find the ratio of the areas.

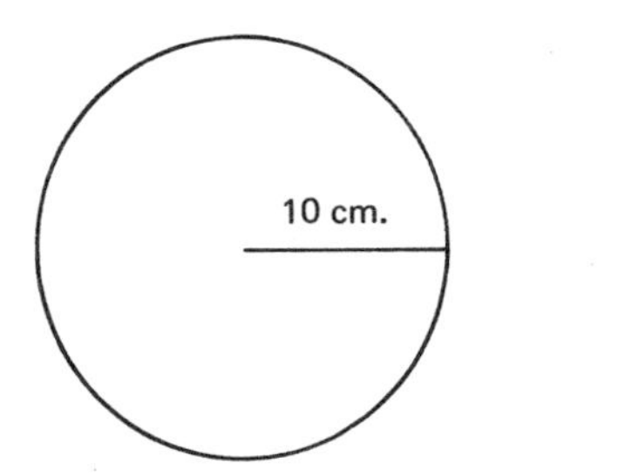


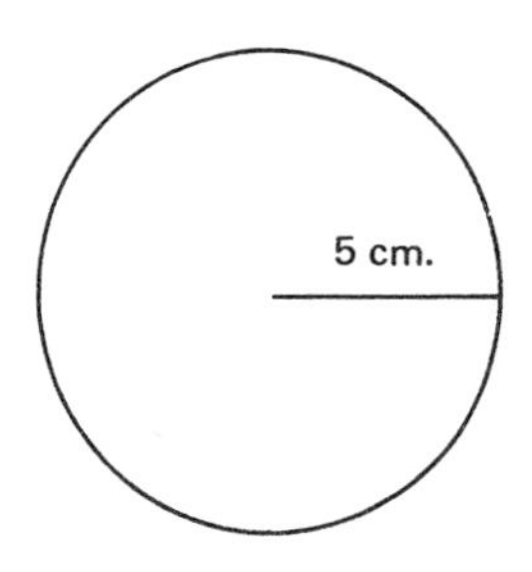


FIGURE 1.85

SOLUTION. (a) The ratio of similarity is given by

$$\frac{r}{r'} = \frac{\overset{2}{\cancel{10}}\ \cancel{\text{cm.}}}{\underset{1}{\cancel{5}}\ \cancel{\text{cm.}}}$$

$$= 2$$

(b) The ratio of the areas is given by

$$\left(\frac{r}{r'}\right)^2 = 2^2$$

$$= 4$$

## CLASS EXERCISES

4. (a) Find the ratio of similarity of two squares with side lengths 15 inches and 3 inches.

   (b) Find the ratio of the areas.

5. (a) Find the ratio of similarity of two circles with radii 24 centimeters and 4 centimeters.

   (b) Find the ratio of the areas.

### TRIANGLES, PARALLELOGRAMS, AND TRAPEZOIDS

Let b be a base and h be the corresponding height of one triangle (or parallelogram), and let $b'$ be a base and $h'$ be the corresponding height of another triangle (or parallelogram). Then the two triangles (or parallelograms) are similar if

$$\frac{b}{b'} = \frac{h}{h'}.$$

This statement is true of two trapezoids if b and $b'$ denote the longer bases of each.

*Two triangles are similar if the lengths of their corresponding sides are proportional.*

In Figure 1.86, the triangles are similar. Corresponding sides are marked with the same number of bars. The sides of length a, b, c of one triangle correspond, respectively, to the sides of length a′, b′, c′ of the other. Assume a > a′.

$$\frac{a}{a'} = \frac{b}{b'} = \frac{c}{c'} = \textit{ratio of similarity}$$

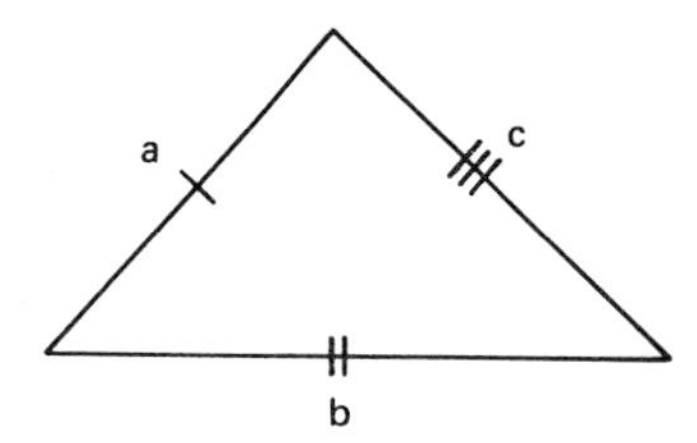


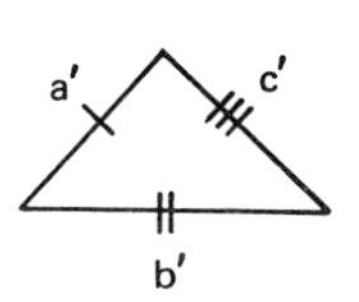


FIGURE 1.86 Similar triangles.

EXAMPLE 4. Suppose the triangles in Figure 1.87 are similar.

(a) Find b′. (b) What is the ratio of similarity? (c) Find c′.

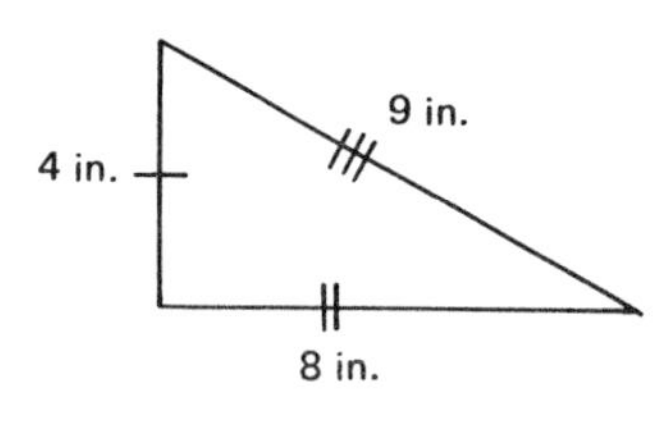


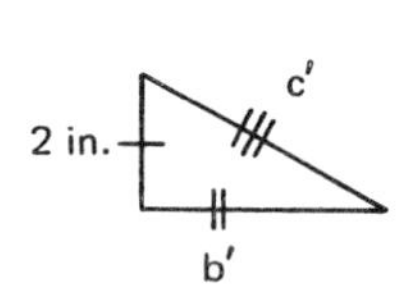


FIGURE 1.87

SOLUTION.

(a)

$$\frac{a}{a'} = \frac{b}{b'} = \frac{c}{c'}$$

$$\frac{4 \text{ in.}}{2 \text{ in.}} = \frac{8 \text{ in.}}{b'} = \frac{9 \text{ in.}}{c'}$$

Solve for b′: First divide numerator and denominator on the left by 2 in.

$$\frac{\overset{2}{\cancel{4 \text{ in.}}}}{\underset{1}{\cancel{2 \text{ in.}}}} = \frac{8 \text{ in.}}{b'} \qquad \text{Now cross-multiply.}$$

$$2b' = 8 \text{ in.}$$

$$\frac{2b'}{2} = \frac{8 \text{ in.}}{2}$$

$$b' = 4 \text{ in.}$$

(b) The ratio of similarity is 2:

$$\frac{a}{a'} = \frac{4 \text{ in.}}{2 \text{ in.}} = 2$$

(c) Use part (b) to find $c'$.

$$\frac{a}{a'} = \frac{c}{c'}$$

$$2 = \frac{9 \text{ in.}}{c'} \qquad \text{Cross-multiply.}$$

$$2c' = 9 \text{ in.}$$

$$\frac{2c'}{2} = \frac{9 \text{ in.}}{2}$$

$$c' = \frac{9}{2} \text{ in. or } 4\frac{1}{2} \text{ in.}$$

*When two parallelograms are similar, the lengths of their corresponding sides are proportional. And when two trapezoids are similar, the lengths of their corresponding sides are proportional.*

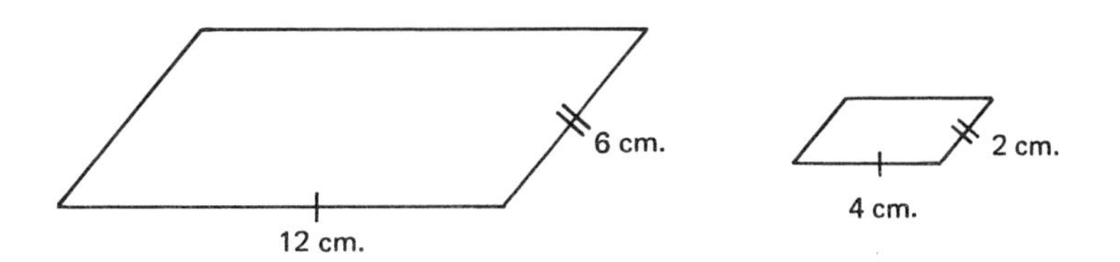


FIGURE 1.88 Similar parallelograms.

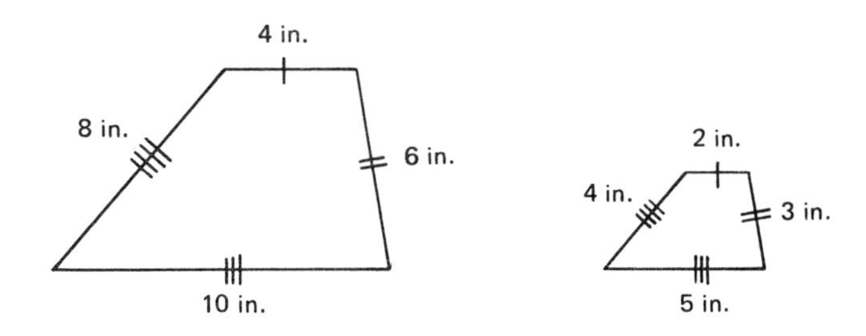


FIGURE 1.89 Similar trapezoids.

When two triangles are similar or two parallelograms are similar, let b and $b'$ be corresponding bases, $b > b'$, let h be the height corresponding to base b and let $h'$ be the height corresponding to base $b'$. Then

$$\frac{b}{b'} = \frac{h}{h'} = \textit{ratio of similarity}$$

When two trapezoids are similar, let $b_1$ and $b_1'$ be one pair of corresponding bases, $b_1 > b_1'$, let $b_2$ and $b_2'$ be the other pair of corresponding bases, and let h and $h'$ be the corresponding heights. Then

$$\frac{b_1}{b_1'} = \frac{b_2}{b_2'} = \frac{h}{h'} = \textit{ratio of similarity}$$

*For similar triangles, similar parallelograms, and similar trapezoids, if* R *is the ratio of similarity, then the ratio of the areas is* $R^2$.

EXAMPLE 5. Consider the given parallelograms. (a) Find h. (b) Find the ratio of the areas.

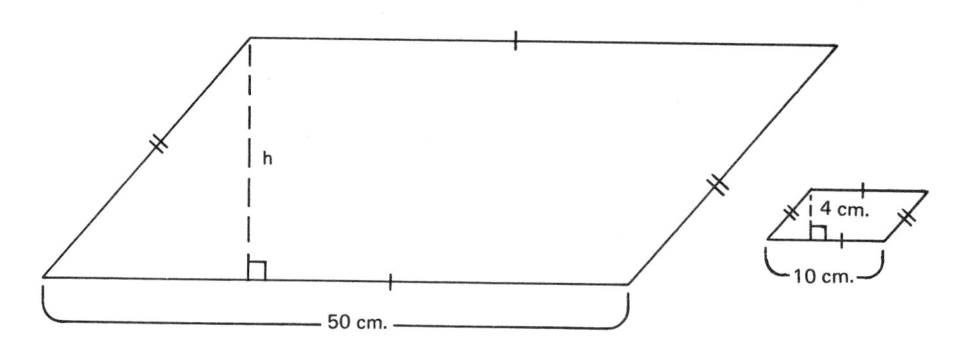


FIGURE 1.90

SOLUTION.

(a)

$$\frac{b}{b'} = \frac{h}{h'}$$

$$\frac{\overset{5}{\cancel{50\text{ cm.}}}}{\underset{1}{\cancel{10\text{ cm.}}}} = \frac{h}{4\text{ cm.}}$$

$$5 \times 4\text{ cm.} = h$$

$$20\text{ cm.} = h$$

(b) The ratio of similarity is $\frac{50\text{ cm.}}{10\text{ cm.}}$, or 5.

The ratio of the areas is $5^2$, or 25.

## CLASS EXERCISES

6. Assume the triangles in Figure 1.91 are similar.

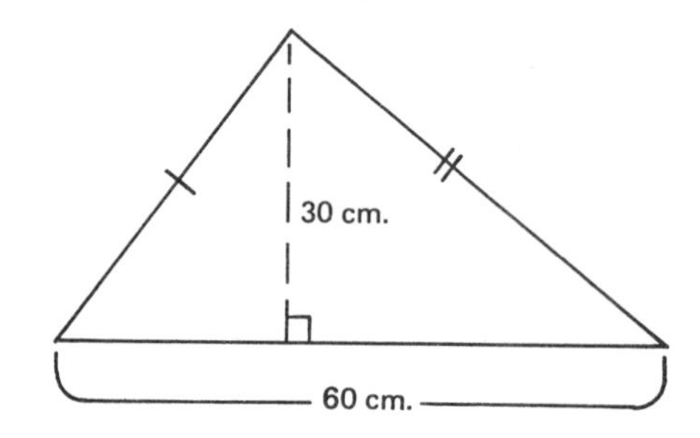


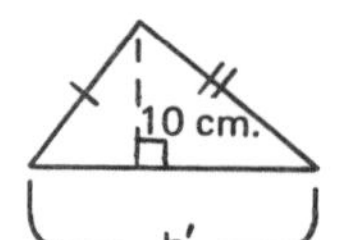


FIGURE 1.91

(a) Find $b'$. (b) What is the ratio of similarity? (c) What is the ratio of the areas?

7. Assume the parallelograms in Figure 1.92 are similar.

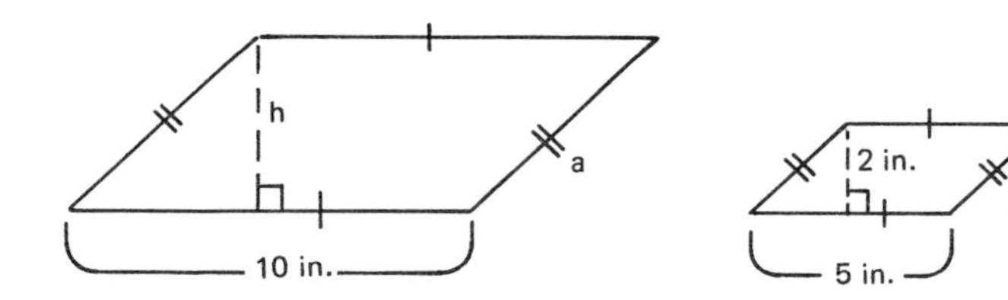


FIGURE 1.92

(a) Find a. (b) What is the ratio of similarity? (c) Find h. (d) What is the ratio of the areas?

8. Assume that the trapezoids in Figure 1.93 are similar.

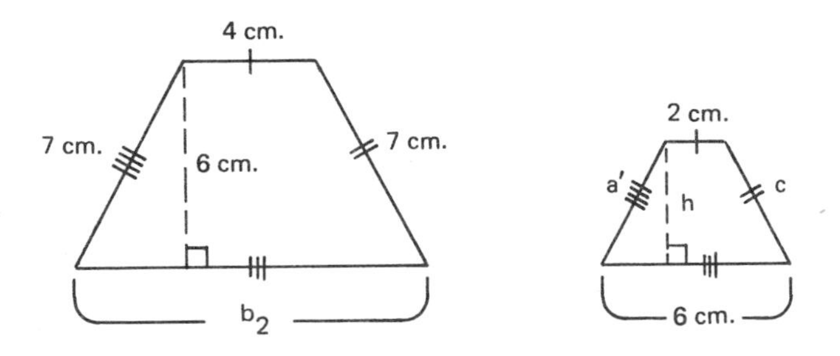


FIGURE 1.93

(a) Find a′. (b) What is the ratio of similarity? (c) Find $b_2$. (d) Find c′.
(e) Find h′. (f) What is the ratio of the areas?

## SOLUTIONS TO CLASS EXERCISES

1. (a) $\frac{\ell}{\ell'} = \frac{w}{w'}$

$$\frac{\overset{4}{\cancel{16}}\ \cancel{\text{cm.}}}{\underset{1}{\cancel{4}}\ \cancel{\text{cm.}}} \overset{?}{=} \frac{\overset{4}{\cancel{12}}\ \cancel{\text{cm.}}}{\underset{1}{\cancel{3}}\ \cancel{\text{cm.}}}$$

$$4 \overset{\checkmark}{=} 4$$

(b) 4

2. $\frac{\ell}{\ell'} = \frac{w}{w'}$

$$\frac{8\ \cancel{\text{in.}}}{5\ \cancel{\text{in.}}} \overset{?}{=} \frac{\overset{2}{\cancel{4}}\ \cancel{\text{in.}}}{\underset{1}{\cancel{2}}\ \cancel{\text{in.}}}$$

$$\frac{8}{5} \overset{\times}{=} 2$$

3. (a) $\frac{\ell}{\ell'} = \frac{w}{w'}$

$\frac{\overset{4}{\cancel{36\text{ cm.}}}}{\underset{1}{\cancel{9\text{ cm.}}}} = \frac{w}{4\text{ cm.}}$

$4 \times 4\text{ cm.} = w$

$16\text{ cm.} = w$

(b) Ratio of similarity $= \frac{\ell}{\ell'}$

$= \frac{36\text{ cm.}}{9\text{ cm.}}$

$= 4$

(c) Ratio of areas $= 4^2$

$= 16$

4. (a) Ratio of similarity $= \frac{15\text{ in.}}{3\text{ in.}} = 5$

(b) Ratio of area $= 5^2 = 25$

5. (a) Ratio of similarity $= \frac{24\text{ cm.}}{4\text{ cm.}} = 6$

(b) Ratio of areas $= 6^2 = 36$

6. (a) $\frac{b}{b'} = \frac{h}{h'}$

$\frac{60\text{ cm.}}{b'} = \frac{\overset{3}{\cancel{30\text{ cm.}}}}{\underset{1}{\cancel{10\text{ cm.}}}}$ Cross-multiply.

$60\text{ cm.} = 3b'$

$\frac{60\text{ cm.}}{3} = \frac{3b'}{3}$

$20\text{ cm.} = b'$

(b) Ratio of similarity $= \frac{h}{h'}$

$= \frac{30\text{ cm.}}{10\text{ cm.}}$

$= 3$

(c) Ratio of areas $= 3^2 = 9$

7. (a) $\frac{a}{a'} = \frac{b}{b'}$

$\frac{a}{3\text{ in.}} = \frac{\overset{2}{\cancel{10\text{ in.}}}}{\underset{1}{\cancel{5\text{ in.}}}}$

$a = 2 \times 3\text{ in.}$

$a = 6\text{ in.}$

(b) Ratio of similarity $= \frac{b}{b'}$

$= \frac{10\text{ in.}}{5\text{ in.}}$

$= 2$

(c) $\frac{b}{b'} = \frac{h}{h'}$

$2 = \frac{h}{2\text{ in.}}$

$4\text{ in.} = h$

(d) Ratio of areas = (Ratio of similarity)$^2$

$= 2^2$

$= 4$

8\. (a)

$$\frac{\overset{2}{\cancel{4}\ \text{cm.}}}{\underset{1}{\cancel{2}\ \text{cm.}}} = \frac{7\ \text{cm.}}{a'}$$

$$2a' = 7\ \text{cm.}$$

$$\frac{2a'}{2} = \frac{7\ \text{cm.}}{2}$$

$$a' = \frac{7}{2}\ \text{cm.}$$

(b) Ratio of similarity $= \dfrac{4\ \text{cm.}}{2\ \text{cm.}} = 2$

(c)

$$2 = \frac{b_2}{6\ \text{cm.}}$$

$$12\ \text{cm.} = b_2$$

(d)

$$\frac{7\ \text{cm.}}{\frac{7}{2}\ \text{cm.}} = \frac{7\ \text{cm.}}{c'}$$

The numerators are equal; the denominators must also be equal.

$$\frac{7}{2}\ \text{cm.} = c'$$

(e)

$$2 = \frac{6\ \text{cm.}}{h'}$$

$$2h' = 6\ \text{cm.}$$

$$\frac{2h'}{2} = \frac{6\ \text{cm.}}{2}$$

$$h' = 3\ \text{cm.}$$

(f) Ratio of areas = (Ratio of similarity)$^2$ = $2^2 = 4$

NAME

CLASS

ANSWERS

## HOME EXERCISES

In Exercises 1-4, determine whether the given rectangles are similar.

1.

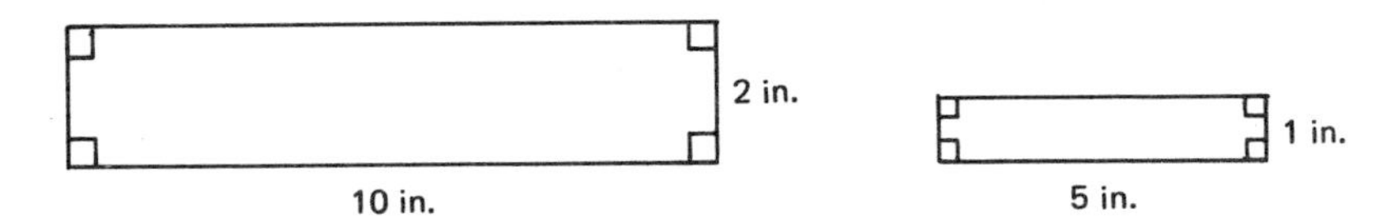


FIGURE 1.94

2.

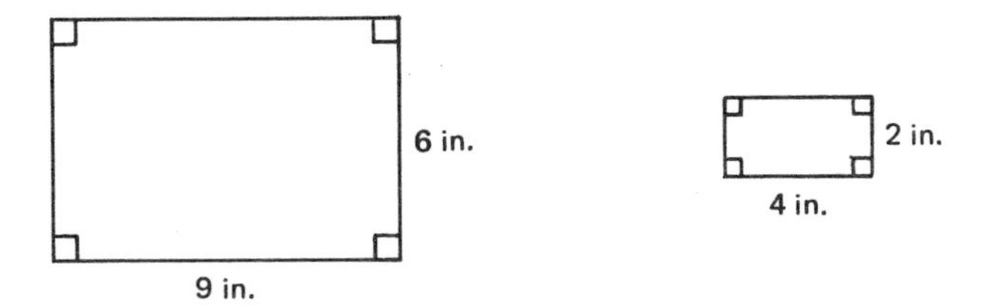


FIGURE 1.95

3.

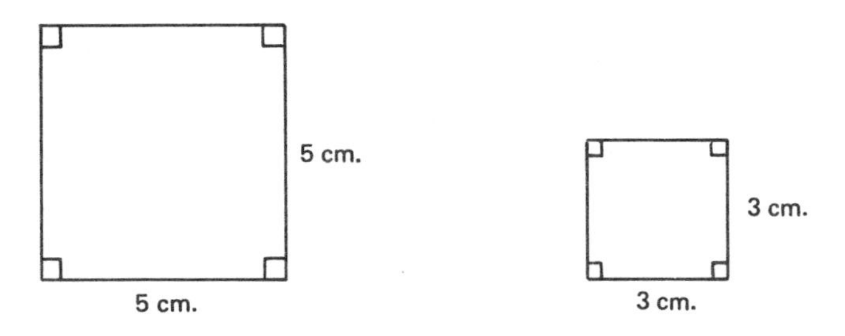


FIGURE 1.96

4.

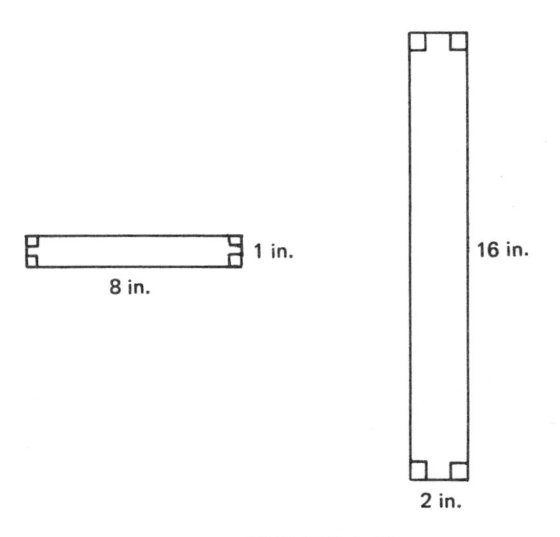


FIGURE 1.97

1. ______

2. ______

3. ______

4. ______

ANSWERS

5. (a) ______

(b) ______

(c) ______

6. (a) ______

(b) ______

(c) ______

7. (a) ______

(b) ______

(c) ______

8. (a) ______

(b) ______

(c) ______

9. ______

10. ______

In Exercises 5-8, suppose the given rectangles are similar.

(a) Find $w'$ or $\ell'$. (b) What is the ratio of similarity?

(c) What is the ratio of the areas?

5.

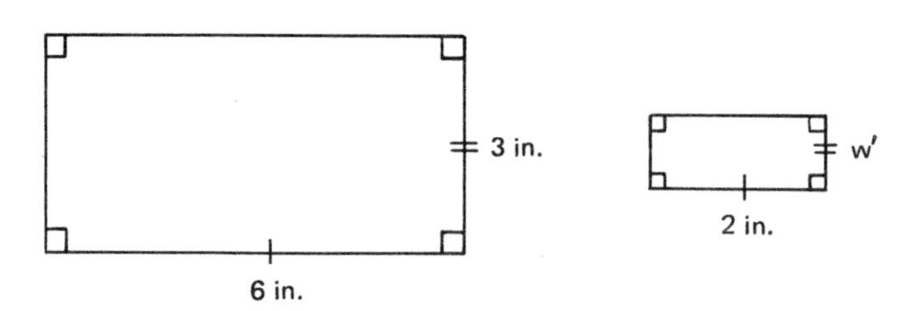


FIGURE 1.98

6.

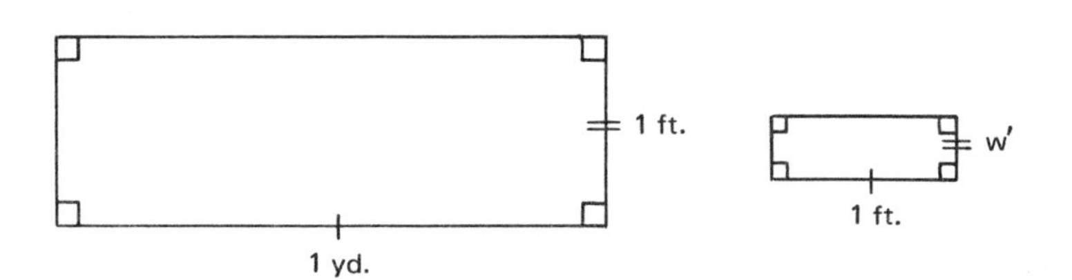


FIGURE 1.99

7.

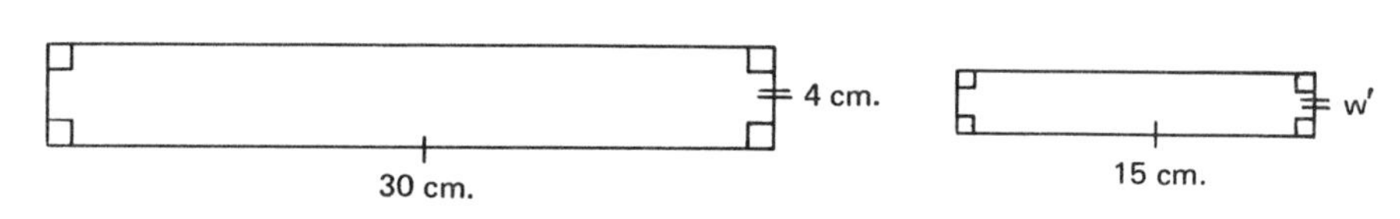


FIGURE 1.100

8.

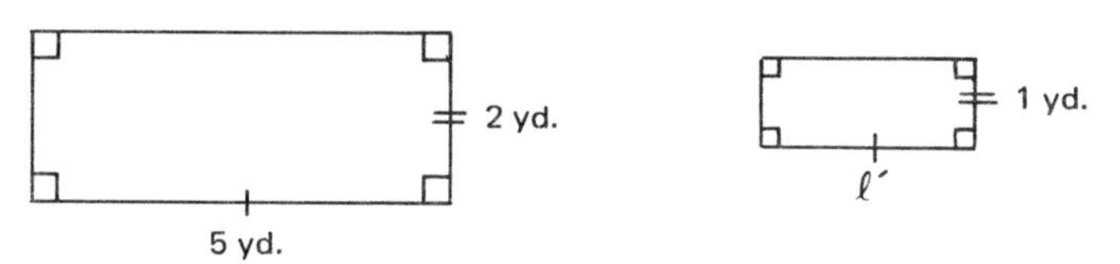


FIGURE 1.101

9. Suppose the ratio of similarity of two similar rectangles is 7. If the length of the larger one is 21 inches, find the length of the smaller one.

10. Suppose the ratio of the areas of two similar rectangles is 4. If the width of the larger one is 21 centimeters, find the width of the smaller one.

NAME

CLASS

ANSWERS

11. (a)

(b)

12. (a)

(b)

13.

14.

15. (a)

(b)

In Exercises 11 and 12: (a) Find the ratio of similarity of the given squares. (b) Find the ratio of the areas.

11.

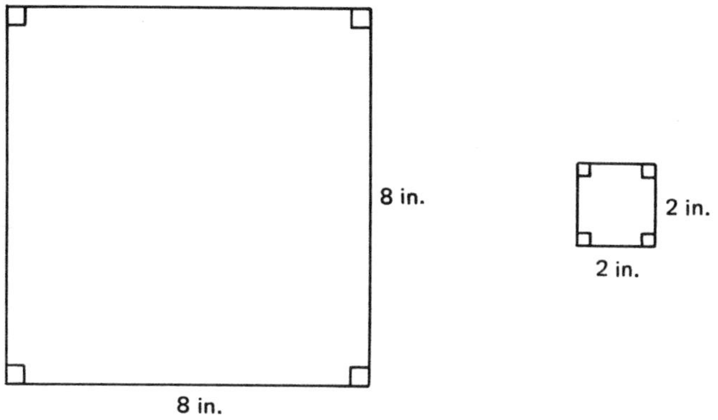

8 in.

8 in.

2 in.

2 in.

FIGURE 1.102

12.

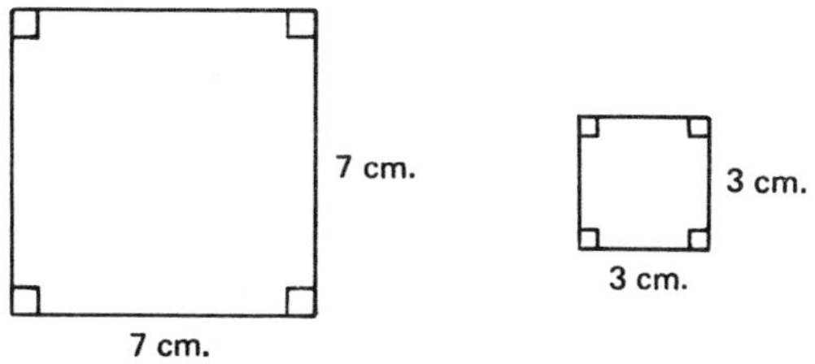

7 cm.

7 cm.

3 cm.

3 cm.

FIGURE 1.103

13. Suppose the ratio of similarity of two squares is 4. If the smaller square has side length 3 inches, find the side length of the larger square.

14. Suppose the ratio of similarity of two squares is 5. If the area of the larger square is 100 square centimeters, find the area of the smaller one.

In Exercises 15 and 16: (a) Find the ratio of similarity of the given circles. (b) Find the ratio of the areas.

15.

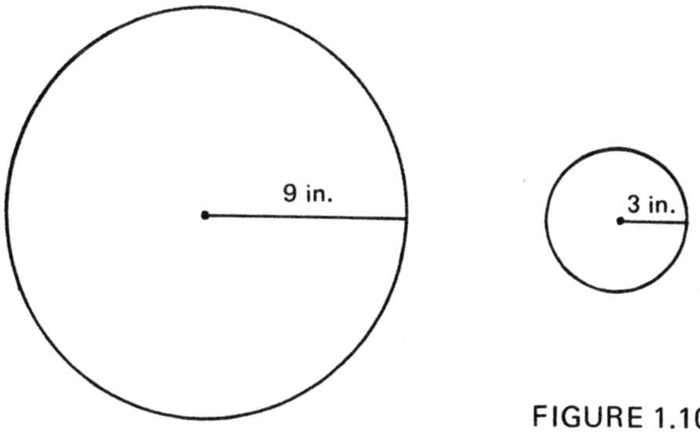

9 in.

3 in.

FIGURE 1.104

ANSWERS

16. (a)________

(b)________

17. ________

18. ________

19. (a)________

(b)________

(c)________

(d)________

(e)________

20. (a)________

(b)________

(c)________

(d)________

(e)________

16.

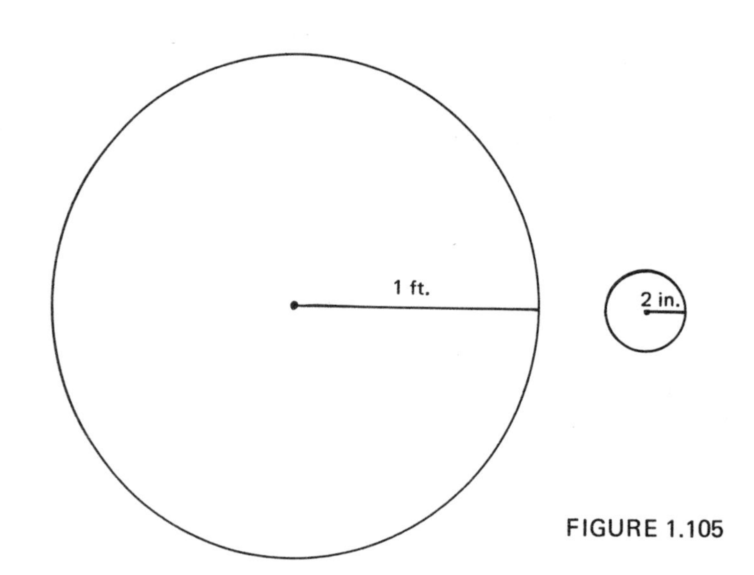


FIGURE 1.105

17. Suppose one circle has radius 8 inches and another circle has area $16\pi$ square inches. What is the ratio of similarity?

18. Suppose one circle has area $100\pi$ square centimeters and another circle has area $4\pi$ square centimeters. What is the ratio of similarity?

In Exercises 19 and 20, assume each pair of triangles is similar.

(a) Find c′. (b) What is the ratio of similarity? (c) Find b.
(d) Find h. (e) What is the ratio of the areas?

19.

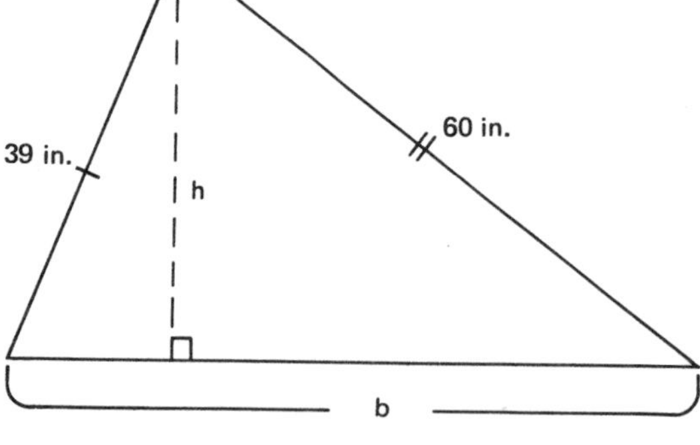


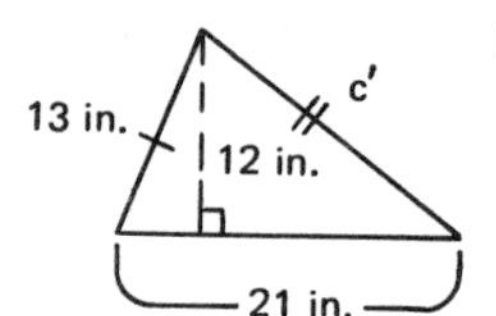


FIGURE 1.106

20.

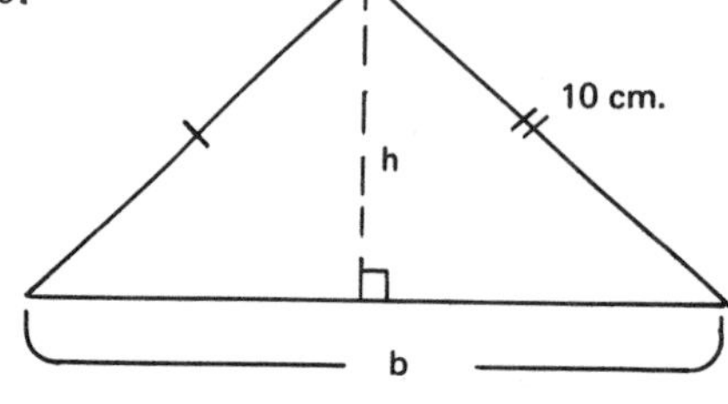


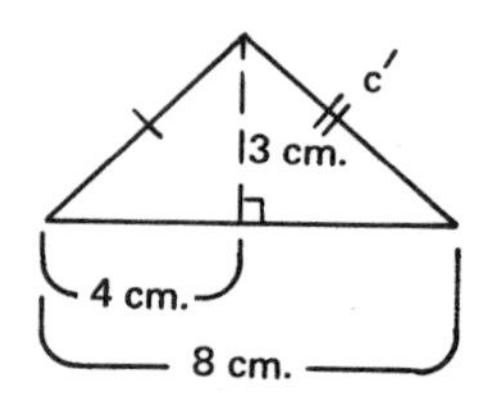


FIGURE 1.107

NAME

CLASS

ANSWERS

In Exercises 21 and 22, assume the parallelograms in each Figure are similar. (a) Find a. (b) What is the ratio of similarity? (c) Find h′. (d) What is the ratio of the areas?

21.

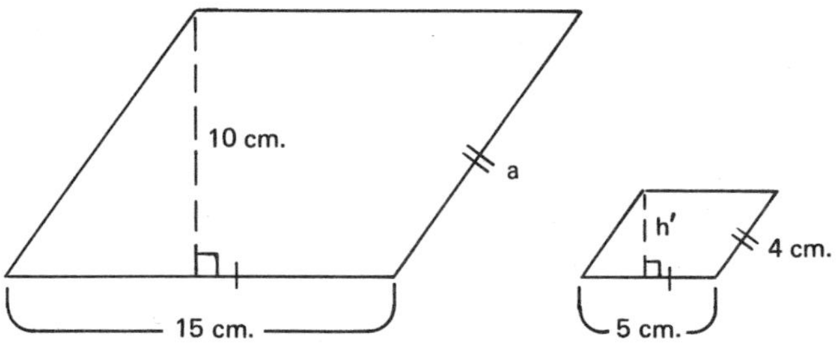

10 cm.
a
15 cm.
h′
4 cm.
5 cm.

FIGURE 1.108

22.

18 in.
a
1 yd.
h′
8 in.
1 ft.

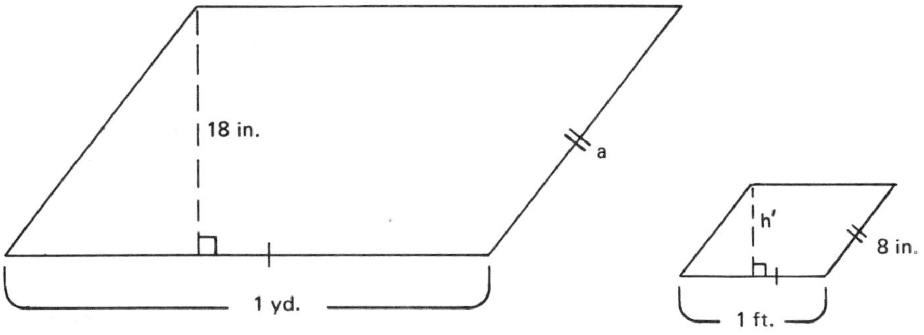

FIGURE 1.109

In Exercises 23 and 24, assume the trapezoids in each Figure are similar. (a) Find a′. (b) What is the ratio of similarity? (c) Find $b_2$. (d) Find c′. (e) Find h. (f) What is the ratio of the areas?

23.

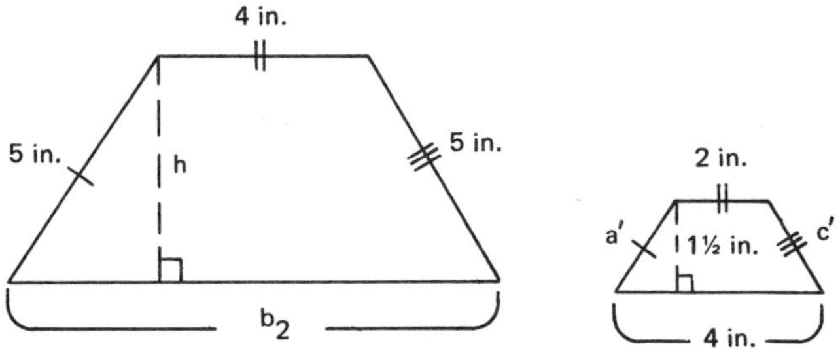

4 in.
5 in.
h
5 in.
$b_2$
2 in.
a′
1½ in.
c′
4 in.

FIGURE 1.110

24.

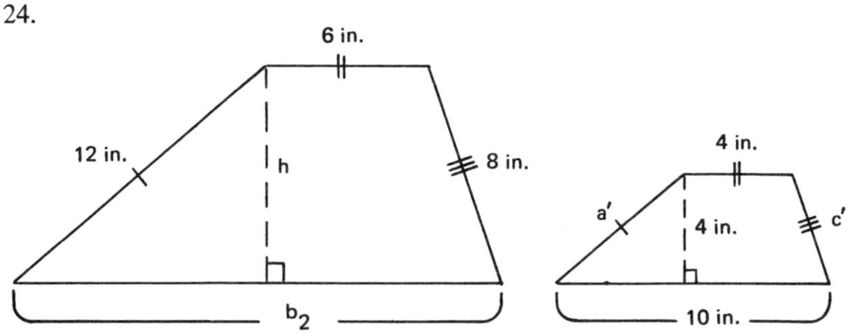

6 in.
12 in.
h
8 in.
$b_2$
4 in.
a′
4 in.
c′
10 in.

FIGURE 1.111

21. (a) ______
(b) ______
(c) ______
(d) ______

22. (a) ______
(b) ______
(c) ______
(d) ______

23. (a) ______
(b) ______
(c) ______
(d) ______
(e) ______
(f) ______

24. (a) ______
(b) ______
(c) ______
(d) ______
(e) ______
(f) ______

ANSWERS

25. ______________

26. ______________

27. ______________

28. ______________

29. ______________

30. ______________

31. ______________

32. ______________

25. Suppose the ratio of similarity of two similar triangles is 4. If the height of the larger triangle is 20 centimeters, find the corresponding height of the smaller one.

26. Suppose the ratio of similarity of two similar triangles is 3. If the smaller triangle has base 4 inches and height 6 inches, find the area of the larger triangle.

27. Suppose the ratio of similarity of two similar parallelograms is 2. If the base of the smaller parallelogram is 12 millimeters, find the corresponding base of the larger one.

28. Suppose the ratio of the areas of two similar parallelograms is 9. If the height of the larger parallelogram is 6 centimeters, find the corresponding height of the smaller one.

29. Suppose the ratio of similarity of two similar trapezoids is 5. If the height of the larger parallelogram is 15 kilometers, find the height of the smaller one.

30. Suppose the ratio of the areas of two similar trapezoids is 4, the area of the larger trapezoid is 96 square inches, and the height and one base of the smaller trapezoid are each 4 inches. Find the other base of the smaller trapezoid.

31. The width and length of a 3 × 5 in. (rectangular) photograph are enlarged proportionately. If the length of the enlargement is 10 inches, what is its width?

32. A 6-foot man casts a 4-foot shadow. How long is the shadow cast by his 4-foot son?

*Let's Review Chapter 1.*

1.1 Length: American and Metric Systems

In Exercises 1-4, convert from one unit to another, as indicated.

1. 3½ feet into inches

2. 5.67 meters into centimeters

3. 10 meters into inches (Express your result to the nearest tenth of an inch.)

4. 32 miles into kilometers (Express your result to the nearest kilometer.)

1.2 Rectangles

5. Find the perimeter of a rectangle that has length 5 centimeters and width 4 centimeters.

6. Find the perimeter of a square whose side length is 9 inches.

7. Find the area of a rectangle with base 9 centimeters and width 7 centimeters.

8. Find the area of a square with side length 4 decimeters.

9. Find the height of a rectangle with base 6 inches and area 42 square inches.

10. Find the area of the indicated figure, which consists of a rectangle and four squares.

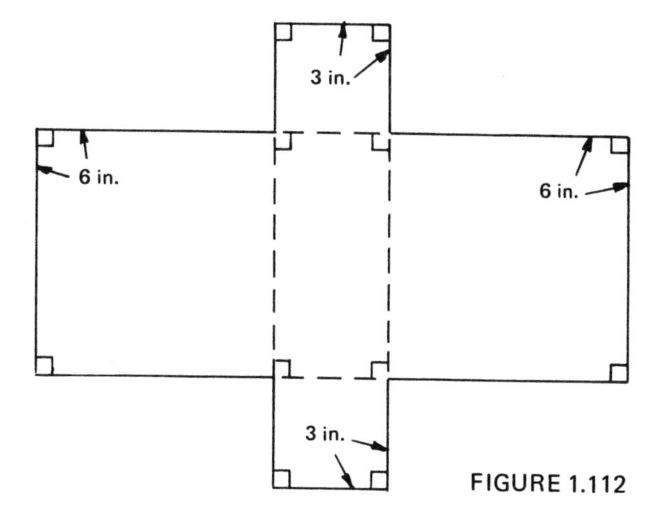


FIGURE 1.112

1.3 Triangles, Parallelograms, Trapezoids

11. Find the area of a triangle with base 9 centimeters and height 10 centimeters.

12. Find the area of a parallelogram with base 7 inches and height 6 inches.

13. Find the base of a parallelogram with height 9 meters and area 108 square meters.

14. Find the area of a trapezoid with bases 8 inches and 10 inches and height 6 inches.

1.4 Circles

15. Find the radius of a circle with diameter 9 centimeters.

16. Find the area of a circle with radius 2 feet.

    (a) Express your answer in terms of $\pi$. (b) Approximate to the nearest tenth of a square foot.

17. Suppose the circumference of a circle is $8\pi$ centimeters. Find (a) the diameter, (b) the radius, (c) the area.

1.5 Similar Figures

18. In Figure 1.113, suppose the given rectangles are similar.
    (a) Find w. (b) What is the ratio of similarity? (c) What is the ratio of the areas?

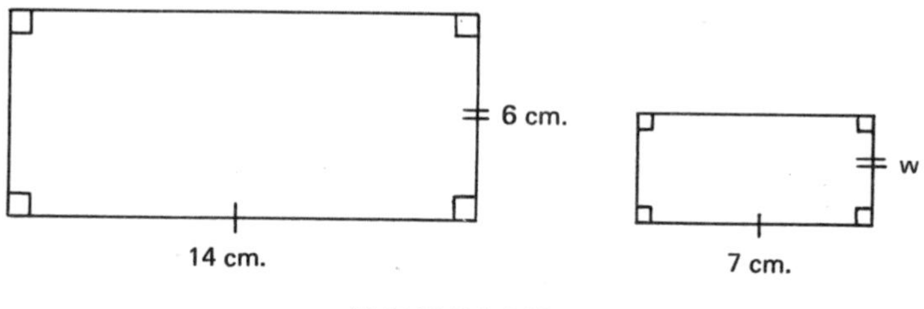


FIGURE 1.113

19. Suppose the ratio of similarity of two squares is 3. If the area of the smaller square is 4 square feet, find the area of the larger square.

20. Suppose the ratio of similarity of two similar triangles is 6. If the height of the larger triangle is 18 inches, find the height of the corresponding smaller triangle.

*Try These Exam Questions for Practice.*

1. Find (a) the perimeter and (b) the area of a rectangle that has length 6 centimeters and width 3 centimeters.

2. Find the base of a triangle with height 8 feet and area 48 square feet.

3. Find the area of a trapezoid with bases 5 inches and 9 inches and height 4 inches.

4. Suppose the diameter of a circle is 6 centimeters. Find (a) the circumference, (b) the radius, (c) the area.

5. Suppose the ratio of similarity of two rectangles is 3. If the smaller rectangle has base 4 inches and height 3 inches, find the area of the larger rectangle.

# Chapter 2
# SOLID GEOMETRY

## DIAGNOSTIC TEST FOR CHAPTER 2

Perhaps you are already familiar with some of the material in Chapter 2. This test will indicate which sections in Chapter 2 you need to study. The question number refers to the corresponding section in Chapter 2. If you answer *all* parts of the question correctly, you may omit the section. But *if any part of your answer is wrong, you should study the section.* When you have completed the test, turn to page 167 for the answers.

2.1 Volume and Surface Area

Suppose a rectangular box has length 2 feet, width 1 foot, and height 3 feet.

(a) Find the volume (in cubic feet).

(b) Find the surface area (in square feet).

(c) Suppose another box has length 6 feet and width 3 feet. What must its height be if the two boxes are similar?

2.2 Spheres, Cylinders, Cones

Express your results in terms of $\pi$.

(a) Suppose the radius of a sphere is 5 inches. Find its volume and surface area.

(b) Suppose a cylinder has base radius 3 centimeters and height 5 centimeters. Find its volume and lateral surface area.

(c) Suppose a cone has base radius 3 inches and height 4 inches. A similar cone has base radius 9 inches. Find the volume of the second cone.

## 2.1 Volume and Surface Area

### RECTANGULAR BOXES

The *volume* of a solid figure, such as a rectangular box, is a measure of how many (cubic) units it can hold.

As you know, the area of a rectangle is the product of its length and width.

Area of rectangle = length × width

Now consider a box that has all rectangular faces, as shown in Figure 2.1. The *volume* of such a *rectangular box*† is the product of its length, width, and height.

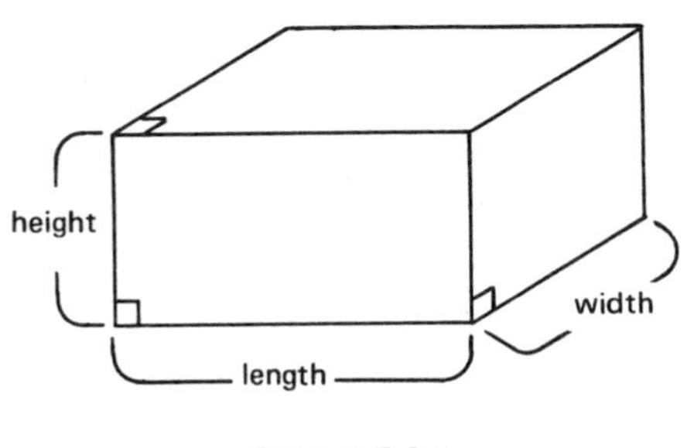


FIGURE 2.1

Volume of rectangular box = length × width × height

In symbols, let

$V$ = volume of a rectangular box,
$\ell$ = length,
$w$ = width,
$h$ = height.

Then

$$V = \ell \cdot w \cdot h$$

The *surface area* of a rectangular box, that is, the sum of the areas of its 6 faces, is obtained by considering Figure 2.1:

| | |
|---|---|
| $2\ell w$ | (top and bottom) |
| $+ 2\ell h$ | (front and back) |
| $+ 2wh$ | (left and right sides) |

Let $S$ = surface area of a rectangular box. Then

$$S = 2\ell w + 2\ell h + 2wh$$

or

$$S = 2(\ell w + \ell h + wh)$$

In Figure 2.2, the length of the box is 4 inches, the width is 3 inches, and the height is 2 inches. The volume is given by:

$$V = \ell \cdot w \cdot h$$

†Specifically, the volume of the region enclosed by the box

$= 4 \text{ inches} \times 3 \text{ inches} \times 2 \text{ inches} = 24 \text{ inches} \times \text{inches} \times \text{inches},$

or 24 cubic inches

The surface area is given by

$S = 2(\ell w + \ell h + wh)$

$= 2(4 \text{ inches} \times 3 \text{ inches} + 4 \text{ inches} \times 2 \text{ inches} + 3 \text{ inches} \times 2 \text{ inches})$

$= 2(12 \text{ square inches} + 8 \text{ square inches} + 6 \text{ square inches})$

$= 2(26 \text{ square inches})$

$= 52 \text{ square inches}$

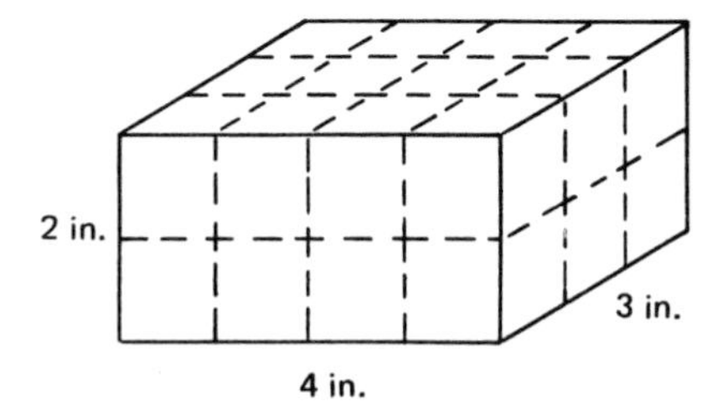


FIGURE 2.2

Note that the unit of volume is cubic inches, whereas the unit of surface area is square inches.

$\text{inches} \times \text{inches} \times \text{inches} = \text{cubic inches}$

Write

$\text{in.}^3$ for cubic inches

$\text{ft.}^3$ for cubic feet

$\text{cm.}^3$ for cubic centimeters, etc.

EXAMPLE 1. Find (a) the volume and (b) the surface area of the rectangular box shown in Figure 2.3.

SOLUTION.

(a)

$V = \ell \cdot w \cdot h$

$= 7 \text{ in.} \times 3 \text{ in.} \times 5 \text{ in.}$

$= 105 \text{ in.}^3$

(b)

$S = 2(\ell w + \ell h + wh)$

$= 2(7 \text{ in.} \times 3 \text{ in.} + 7 \text{ in.} \times 5 \text{ in.} + 3 \text{ in.} \times 5 \text{ in.})$

$= 2(21 \text{ in.}^2 + 35 \text{ in.}^2 + 15 \text{ in.}^2)$

$= 142 \text{ in.}^2$

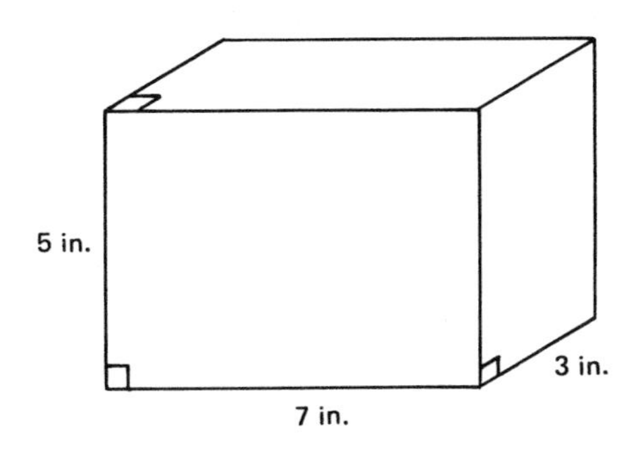


FIGURE 2.3

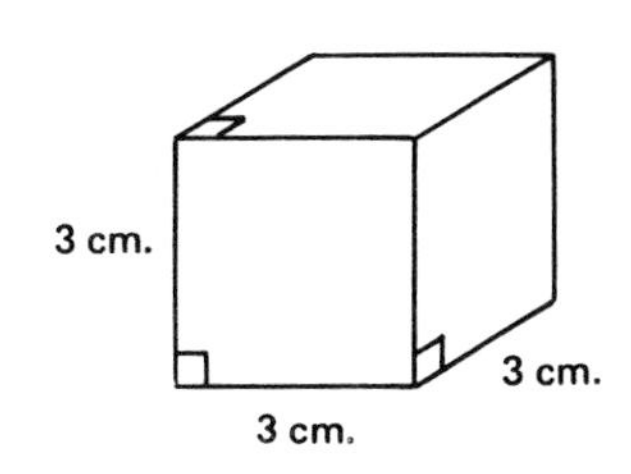


FIGURE 2.4

A *cube* is a rectangular box in which length, width, and height are equal.

$$\text{Volume of cube} = s^3,$$

$$\text{Surface Area of cube} = 2(s^2 + s^2 + s^2) = 6s^2,$$

where s is the length of a side.

EXAMPLE 2. Find (a) the volume and (b) the surface area of a cube if the length of a side is 3 centimeters.

SOLUTION. See Figure 2.4.

(a) $V = s^3$

$= (3 \text{ cm.})^3$

$= 27 \text{ cm.}^3$

(b) $S = 6s^2$

$= 6(3 \text{ cm.})^2$

$= 6 \times 9 \text{ cm.}^2$

$= 54 \text{ cm.}^2$

## CLASS EXERCISES

In Exercises 1-3, find (a) the volume and (b) the surface area of each rectangular box.

1. Length 3 inches, width 1 inch, height 4 inches

2.

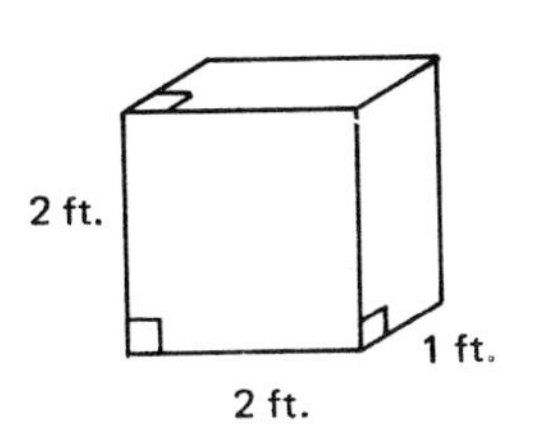


FIGURE 2.5

3.

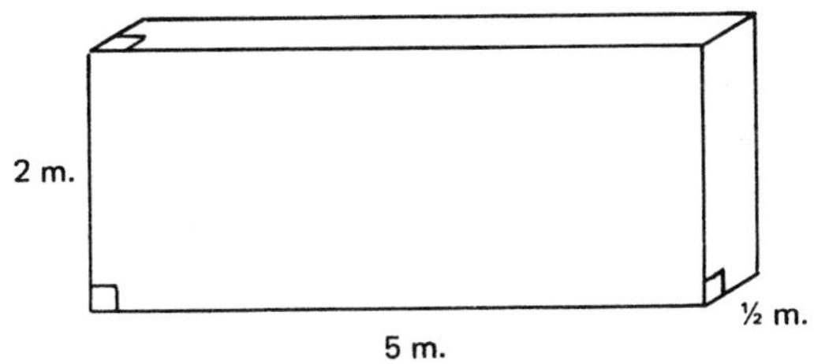


FIGURE 2.6

4. Find (a) the volume and (b) the surface area of a cube if the length of each side is 4 inches.

## SIMILARITY

Two rectangular boxes are *similar*, that is, have the same shape, if corresponding sides are proportional. Let one such box have

length $\ell$, width w, height h;

and let the other have

length $\ell'$, width w′, height h′.

Assume $\ell > \ell'$. Then *the two boxes are similar if*

$$\frac{\ell}{\ell'} = \frac{w}{w'} = \frac{h}{h'}.$$

In this case, $\frac{\ell}{\ell'}$, is called the *ratio of similarity.*

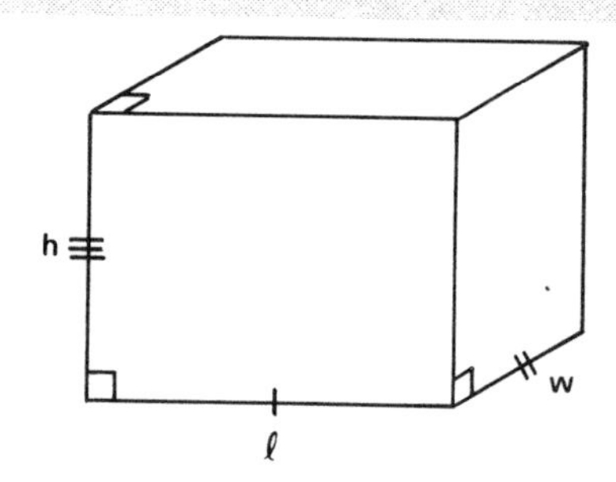


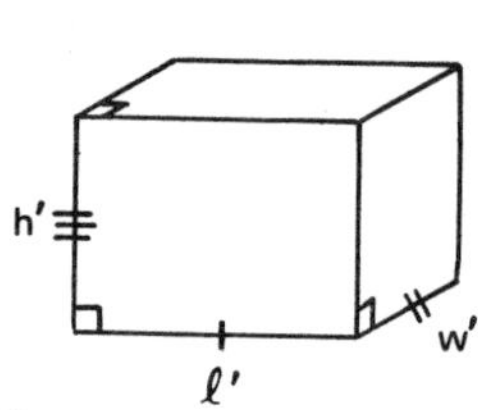


FIGURE 2.7 Similar boxes.

$$\frac{\ell}{\ell'} = \frac{w}{w'} = \frac{h}{h'}$$

EXAMPLE 3. Consider the rectangular boxes in Figure 2.8 (a), (b), and (c).

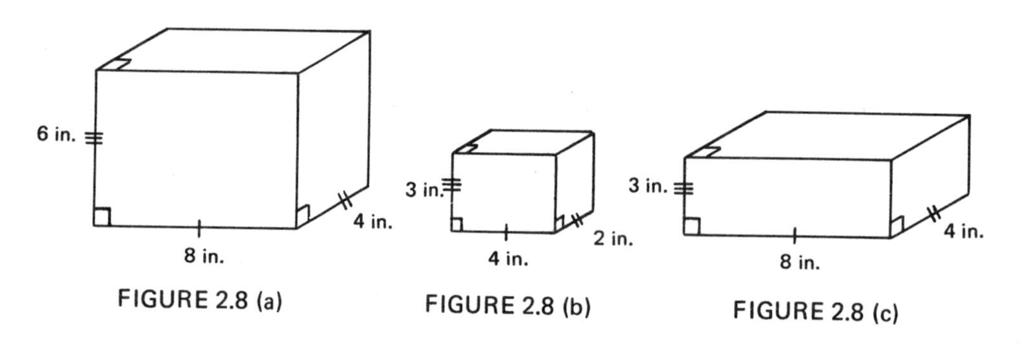


FIGURE 2.8 (a) FIGURE 2.8 (b) FIGURE 2.8 (c)

(a) Boxes (a) and (b) are similar because

$$\frac{\ell}{\ell'} = \frac{w}{w'} = \frac{h}{h'}:$$

$$\frac{8\text{ in.}}{4\text{ in.}} = \frac{4\text{ in.}}{2\text{ in.}} = \frac{6\text{ in.}}{3\text{ in.}} = 2$$

The ratio of similarity is 2.

(b) Boxes (a) and (c) are *not* similar. Although

$$\frac{\ell}{\ell'} = \frac{w}{w'}:$$

$$\frac{8 \text{ in.}}{8 \text{ in.}} = \frac{4 \text{ in.}}{4 \text{ in.}} = 1,$$

the third ratio is *not* 1:

$$\frac{h}{h'} = \frac{6 \text{ in.}}{3 \text{ in.}} = 2$$

*All three ratios,*

$$\frac{\ell}{\ell'}, \frac{w}{w'}, \frac{h}{h'},$$

*must be equal in order for the boxes to be similar.*

(c) Compare the surface areas of the similar boxes (a) and (b).

For box (a):

$$S = 2(\ell w + \ell h + wh)$$

$$= 2(8 \text{ in.} \times 4 \text{ in.} + 8 \text{ in.} \times 6 \text{ in.} + 4 \text{ in.} \times 6 \text{ in.})$$

$$= 2(32 \text{ in.}^2 + 48 \text{ in.}^2 + 24 \text{ in.}^2)$$

$$= 208 \text{ in.}^2$$

For box (b):

$$S' = 2(\ell' \cdot w' + \ell' \cdot h' + w' \cdot h')$$

$$= 2(4 \text{ in.} \times 2 \text{ in.} + 4 \text{ in.} \times 3 \text{ in.} + 2 \text{ in.} \times 3 \text{ in.})$$

$$= 2(8 \text{ in.}^2 + 12 \text{ in.}^2 + 6 \text{ in.}^2)$$

$$= 52 \text{ in.}^2$$

Here, $208 \text{ in.}^2 = 4 \times 52 \text{ in.}^2$. The ratio of similarity is 2 and the ratio of the surface areas is $2^2$, or 4.

In general, *if the ratio of similarity of two rectangular boxes is* R, *then the ratio of the surface areas is* $R^2$.

(d) Compare the volumes of the similar boxes (a) and (b).

For box (a):

$$V = \ell \cdot w \cdot h$$

$$= 8 \text{ in.} \times 4 \text{ in.} \times 6 \text{ in.}$$

$$= 192 \text{ in.}^3$$

For box (b):

$$V' = \ell' \cdot w' \cdot h'$$

$$= 4 \text{ in.} \times 2 \text{ in.} \times 3 \text{ in.}$$

$$= 24 \text{ in.}^3$$

Observe that 192 in.$^3$ = 8 × 24 in.$^3$ The ratio of similarity is 2 and the ratio of the volumes is $2^3$, or 8.

In general, *if the ratio of similarity of two rectangular boxes is* R, *then the ratio of the volumes is* $R^3$.

Any two cubes have the same shape and are therefore similar. In fact, $\ell = w = h$ (and $\ell' = w' = h'$), and therefore,

$$\frac{\ell}{\ell'} = \frac{w}{w'} = \frac{h}{h'}$$

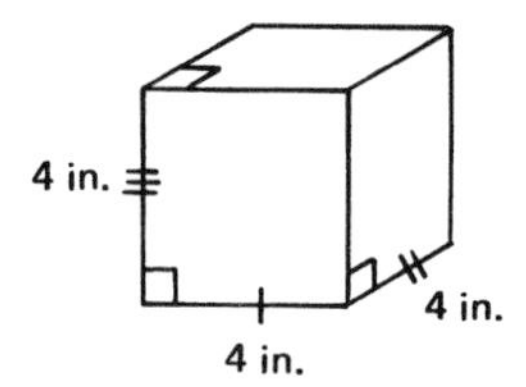


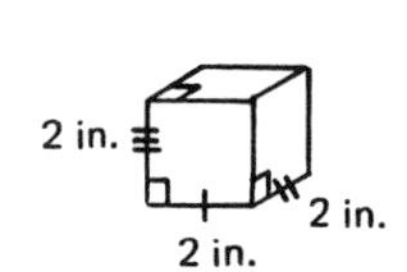


FIGURE 2.9 Any two cubes are similar. Here,

$$\frac{4 \text{ in.}}{2 \text{ in.}} = \frac{4 \text{ in.}}{2 \text{ in.}} = \frac{4 \text{ in.}}{2 \text{ in.}}$$

The ratio of similarity is 2. The ratio of the surface areas is 4. The ratio of the volumes is 8.

## CLASS EXERCISES

5. (a) Show that the rectangular boxes in Figure 2.10 are similar.

   (b) What is the ratio of similarity?

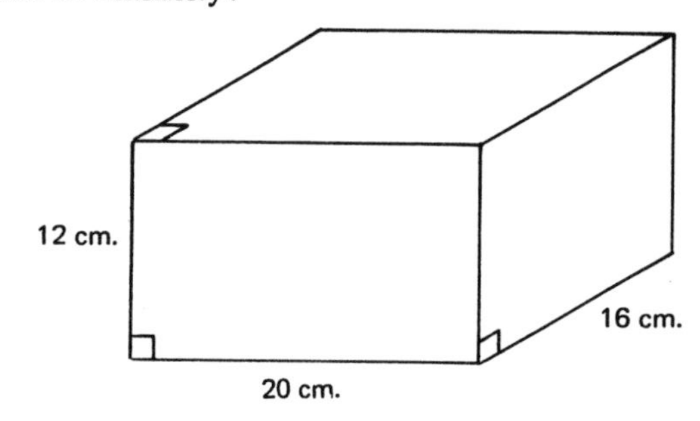


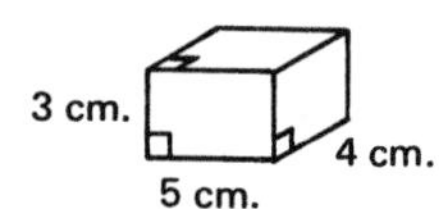


FIGURE 2.10

6. Show that the rectangular boxes in Figure 2.11 are *not* similar.

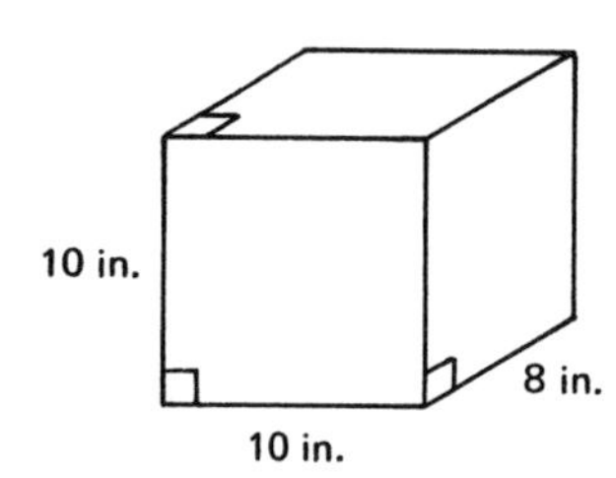


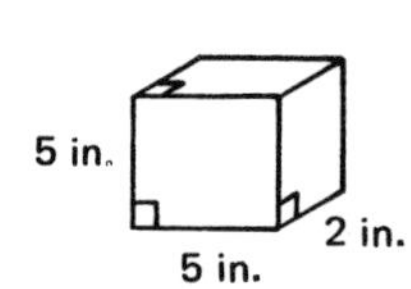


FIGURE 2.11

7. Suppose the rectangular boxes in Figure 2.12 are similar.

   (a) Find w′.

   (b) What is the ratio of similarity?

   (c) Find h.

   (d) What is the ratio of the surface areas?

   (e) What is the ratio of the volumes?

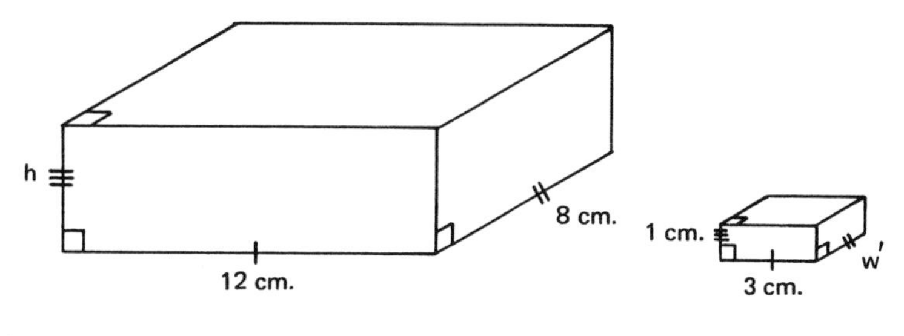


FIGURE 2.12

8. (a) Find the ratio of similarity of two cubes if one has side length 6 inches and the other has side length 2 inches.

   (b) Find the ratio of the surface areas.

   (c) Find the ratio of the volumes.

## SOLUTIONS TO CLASS EXERCISES

1. (a)

$$V = \ell \cdot w \cdot h$$

$$= 3 \text{ in.} \times 1 \text{ in.} \times 4 \text{ in.}$$

$$= 12 \text{ in.}^3$$

(b)

$$S = 2(\ell w + \ell h + wh)$$

$$= 2(3 \text{ in.} \times 1 \text{ in.} + 3 \text{ in.} \times 4 \text{ in.} + 1 \text{ in.} \times 4 \text{ in.})$$

$$= 2(3 \text{ in.}^2 + 12 \text{ in.}^2 + 4 \text{ in.}^2)$$

$$= 38 \text{ in.}^2$$

2. (a)

$$V = \ell \cdot w \cdot h$$
$$= 2 \text{ ft.} \times 1 \text{ ft.} \times 2 \text{ ft.}$$
$$= 4 \text{ ft.}^3$$

(b)

$$S = 2(\ell w + \ell h + wh)$$
$$= 2(2 \text{ ft.} \times 1 \text{ ft.} + 2 \text{ ft.} \times 2 \text{ ft.} + 1 \text{ ft.} \times 2 \text{ ft.})$$
$$= 2(2 \text{ ft.}^2 + 4 \text{ ft.}^2 + 2 \text{ ft.}^2)$$
$$= 16 \text{ ft.}^2$$

3. (a)

$$V = \ell \cdot w \cdot h$$
$$= 5\text{m.} \times \frac{1}{2} \text{ m.} \times 2\text{m.}$$
$$= 5\text{m.}^3$$

(b)

$$S = 2(\ell w + \ell h + wh)$$
$$= 2(5\text{m.} \times \frac{1}{2}\text{m.} + 5\text{m.} \times 2\text{m.} + \frac{1}{2}\text{m.} \times 2\text{m.})$$
$$= 2(\frac{5}{2}\text{ m.}^2 + 10\text{m.}^2 + 1\text{m.}^2)$$
$$= 5\text{m.}^2 + 20\text{m.}^2 + 2\text{m.}^2$$
$$= 27\text{m.}^2$$

4. (a)

$$V = s^3$$
$$= (4 \text{ in.})^3$$
$$= 64 \text{ in.}^3$$

(b)

$$S = 6s^2$$
$$= 6(4 \text{ in.})^2$$
$$= 6 \times 16 \text{ in.}^2$$
$$= 96 \text{ in.}^2$$

5. (a)

$$\frac{\ell}{\ell'} \stackrel{?}{=} \frac{w}{w'} \stackrel{?}{=} \frac{h}{h'}$$

$$\frac{20 \text{ cm.}}{5 \text{ cm.}} \stackrel{?}{=} \frac{16 \text{ cm.}}{4 \text{ cm.}} \stackrel{?}{=} \frac{12 \text{ cm.}}{3 \text{ cm.}}$$

$$4 \stackrel{\checkmark}{=} 4 \stackrel{\checkmark}{=} 4$$

(b) The ratio of similarity is 4.

6. $$\frac{\ell}{\ell'} \neq \frac{w}{w'}$$

$$\frac{10 \text{ in.}}{5 \text{ in.}} \stackrel{?}{=} \frac{8 \text{ in.}}{2 \text{ in.}}$$

$$2 \stackrel{\times}{=} 4$$

7. (a) $$\frac{\ell}{\ell'} = \frac{w}{w'}$$

$$\frac{\overset{4}{\cancel{12} \text{ cm.}}}{\underset{1}{\cancel{3} \text{ cm.}}} = \frac{8 \text{ cm.}}{w'}$$

$$4w' = 8 \text{ cm.}$$

$$\frac{4w'}{4} = \frac{8 \text{ cm.}}{4}$$

$$w' = 2 \text{ cm.}$$

(b) Ratio of similarity $= \frac{\ell}{\ell'}$

$= \frac{12 \text{ cm.}}{3 \text{ cm.}}$

$= 4$

(c) $$\frac{\ell}{\ell'} = \frac{h}{h'}$$

$$4 = \frac{h}{1 \text{ cm.}}$$

$$4 \text{ cm.} = h$$

(d) Ratio of surface areas = (Ratio of similarity)$^2$ = $4^2$ = 16

(e) Ratio of volumes = (Ratio of similarity)$^3$ = $4^3$ = 64

8. (a) Ratio of similarity $= \frac{\ell}{\ell'} = \frac{6 \text{ in.}}{2 \text{ in.}} = 3$

(b) Ratio of surface areas = (Ratio of similarity)$^2$ = $3^2$ = 9

(c) Ratio of volumes = (Ratio of similarity)$^3$ = $3^3$ = 27

## HOME EXERCISES

In Exercises 1-6, find (a) the volume and (b) the surface area of each rectangular box.

1. Length 5 inches, width 2 inches, height 2 inches

2. Length 4 feet, width 2 feet, height 8 feet

3. Length 12 centimeters, width 3 centimeters, height 5 centimeters

4. 5. 6.

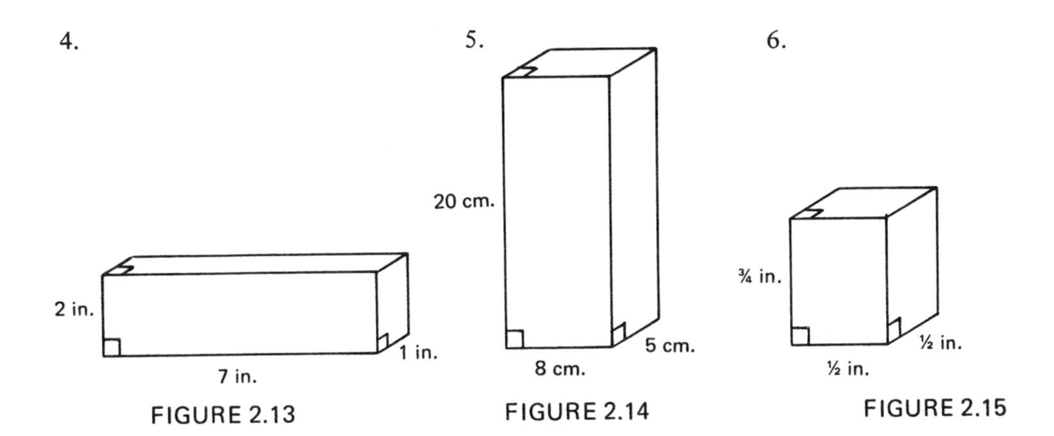


FIGURE 2.13 FIGURE 2.14 FIGURE 2.15

In Exercises 7-9, find (a) the volume and (b) the surface area of each cube.

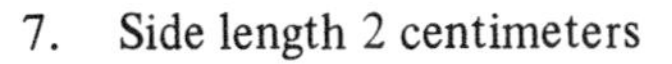

7. Side length 2 centimeters

8. Side length 10 inches

9.

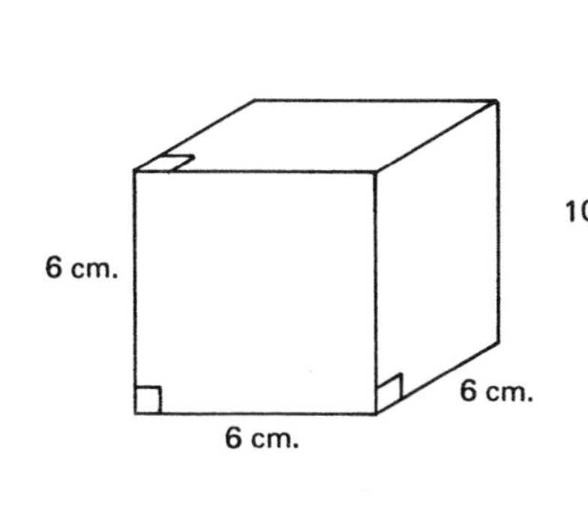


FIGURE 2.16

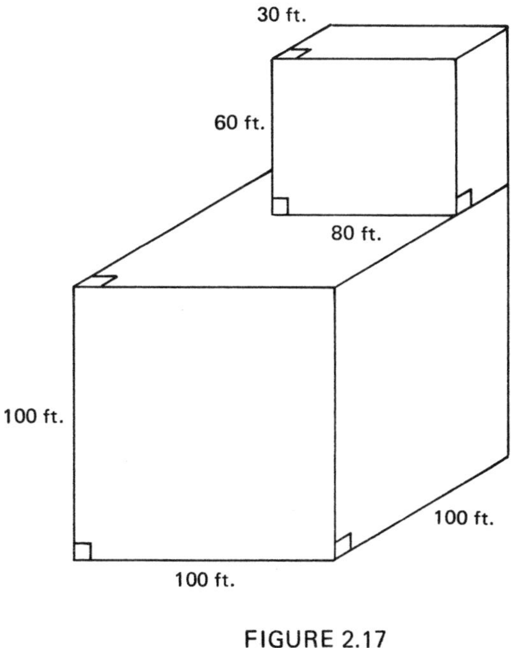


FIGURE 2.17

NAME ______

CLASS ______

ANSWERS

1. (a) ______
   (b) ______
2. (a) ______
   (b) ______
3. (a) ______
   (b) ______
4. (a) ______
   (b) ______
5. (a) ______
   (b) ______
6. (a) ______
   (b) ______
7. (a) ______
   (b) ______
8. (a) ______
   (b) ______
9. (a) ______
   (b) ______

ANSWERS

10. ____________

11. ____________

12. ____________

13. ____________

14. ____________

15. ____________

16. ____________

17. (a) ____________

(b) ____________

18. ____________

19. ____________

10. Find the volume of the solid in Figure 2.17.

11. Find the height of a rectangular box that has length 4 inches, width 2 inches, and volume 40 cubic inches.

12. Find the width of a rectangular box that has length 7 feet, height 10 feet, and volume 140 cubic feet.

13. Find the length of a rectangular box that has width 3 inches, height 8 inches, and volume 96 cubic inches.

14. Find the side length of a cube that has volume 27 cubic inches.

15. Find the volume of air in a room that measures 14 feet by 10 feet by 9 feet.

16. A rectangular swimming pool measures 25 feet by 20 feet by 6 feet. How much water can it hold?

17. (a) What is the volume of a cigar box of length 11 inches, width 6 inches, and height 3 inches?

(b) How much paper is needed to line the inside of the box?

18. A trunk measures 3 feet by 2 feet by 1½ feet. What is its volume?

In Exercises 19-21, determine whether the given rectangular boxes are similar.

19.

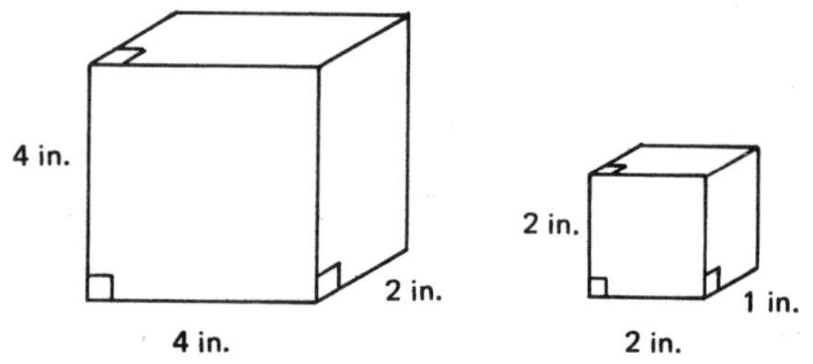

4 in.
2 in.
4 in.
2 in.
1 in.
2 in.

FIGURE 2.18

NAME

CLASS

ANSWERS

20. ______

21. ______

22. (a) ______

(b) ______

(c) ______

(d) ______

(e) ______

23. (a) ______

(b) ______

(c) ______

(d) ______

(e) ______

20.

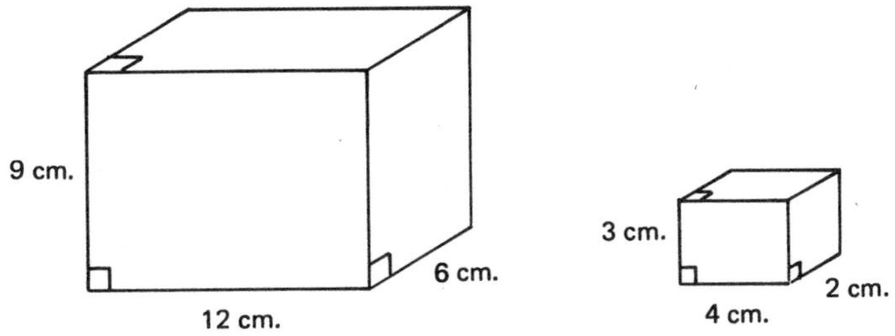

FIGURE 2.19

21.

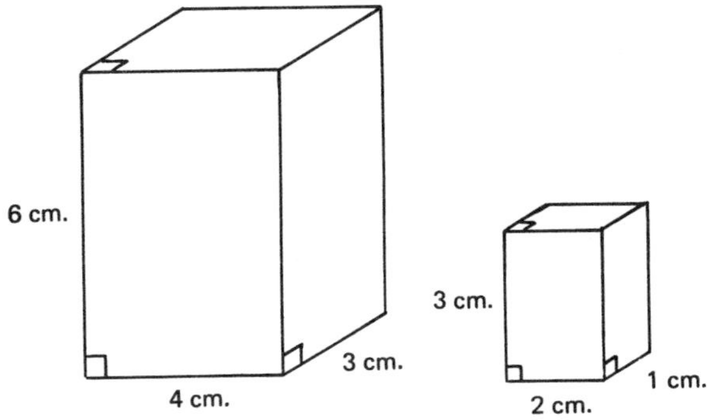

FIGURE 2.20

In Exercises 22-24, suppose the given rectangular boxes are similar.

(a) Find w or w′.

(b) What is the ratio of similarity?

(c) Find h or h′.

(d) What is the ratio of the surface areas?

(e) What is the ratio of the volumes?

22.

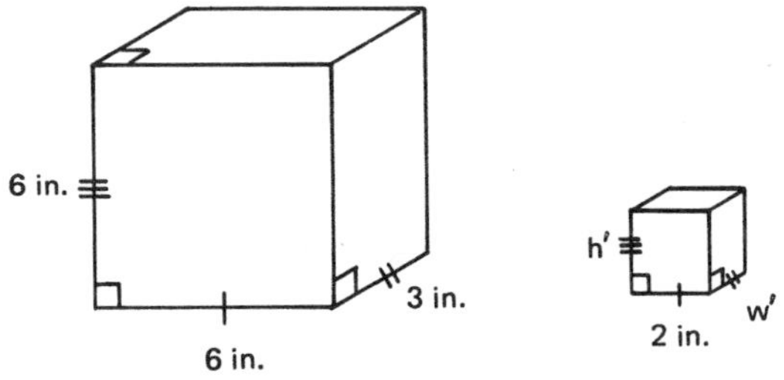

FIGURE 2.21

23.

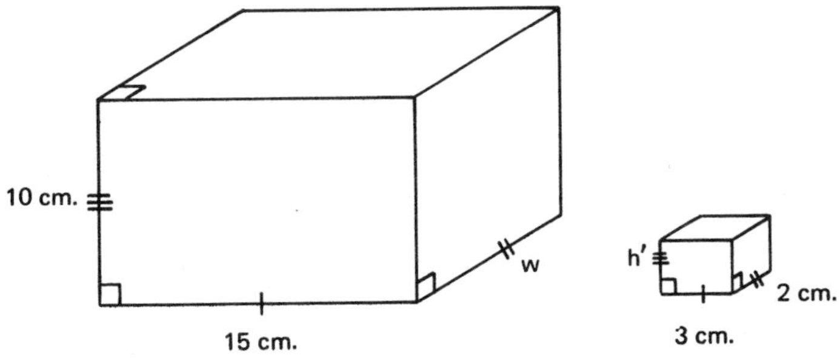

FIGURE 2.22

ANSWERS

24. (a)__________

(b)__________

(c)__________

(d)__________

(e)__________

25.__________

26.__________

27.__________

28. (a)__________

(b)__________

(c)__________

29.__________

30.__________

24.

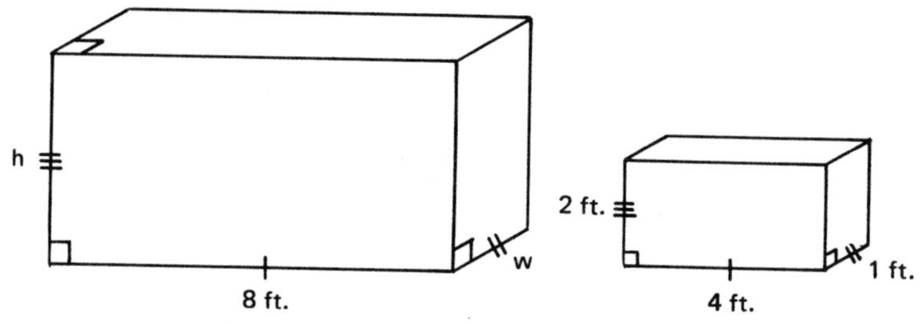

h

w

8 ft.

2 ft.

1 ft.

4 ft.

FIGURE 2.23

25. Suppose the ratio of similarity of two similar rectangular boxes is 6. If the volume of the smaller box is 10 cubic feet, what is the volume of the larger box?

26. Suppose the larger of two similar rectangular boxes has length 8 centimeters. If the ratio of the surface areas is 4, find the length of the smaller box.

27. A rectangular trunk measures 4 feet by 2 feet by 1 foot. A similar trunk has length 2 feet. What are its other dimensions?

28. (a) Find the ratio of similarity of two cubes, one with side length 24 centimeters, the other with side length 6 centimeters.

(b) Find the ratio of the surface areas. (c) Find the ratio of the volumes.

29. Suppose the side length of the smaller of two cubes is 5 inches. If the ratio of similarity is 2, find the volume of the larger cube.

30. Two cubes have volumes of 3000 cubic feet and 3 cubic feet, respectively. Find the ratio of similarity.

## 2.2 Spheres, Cylinders, Cones

### SPHERES

Recall that a circle is a closed figure (in a plane) every point of which is at a *fixed* distance from a given point. We will now consider the "3-dimensional" counterpart of a circle.

DEFINITION. A *sphere* (or ball) is a surface *in space* every point of which is at a fixed distance from a given point, known as the *center* of the sphere. The fixed distance is called the *radius* of the sphere.

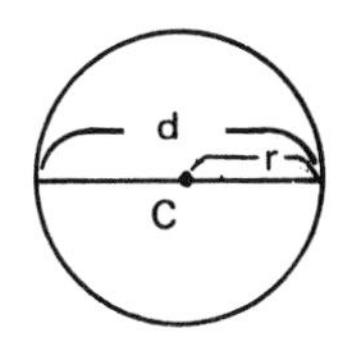


FIGURE 2.24 A circle with center C, radius r, and diameter d.

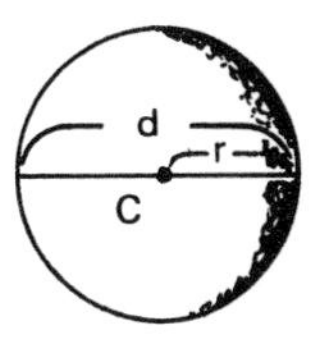


FIGURE 2.25 A sphere with center C, radius r, and diameter d.

The volume, V, of a sphere† is given by the formula

$$V = \frac{4}{3}\pi r^3, \text{ or } V = \frac{4\pi r^3}{3},$$

where r is the radius. The surface area, S, is given by

$$S = 4\pi r^2.$$

EXAMPLE 1. A sphere has radius 3 inches.

(a) Express the volume in terms of $\pi$.

(b) Approximate the volume to the nearest cubic inch.

(c) Express the surface area in terms of $\pi$.

SOLUTION.

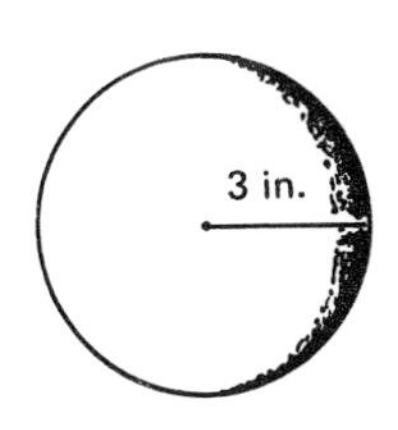


FIGURE 2.26

(a) $V = \frac{4}{3}\pi r^3$

$= \frac{4}{3}\pi(3 \text{ in.})^3$

$= \frac{4}{\cancel{3}_{1}}\pi \, \cancel{27 \text{ in.}^3}^{\,9 \text{ in.}^3}$

$= 36\pi \text{ in.}^3$

†We will speak of the volume of a sphere rather than of a spherical region. Similarly, we will speak of the volume of a cylinder rather than of a cylindrical region, etc.

(b)

$$V \approx \underbrace{36 \times 3.14}_{113.04} \text{ in.}^3$$

$$V \approx 113 \text{ in.}^3$$

(c)

$$S = 4\pi r^2$$

$$= 4\pi(3 \text{ in.})^2$$

$$= 4\pi \times 9 \text{ in.}^2$$

$$= 36\pi \text{ in.}^2$$

## CLASS EXERCISES

Consider a sphere with the given radius.

(a) Express the volume in terms of $\pi$.

(b) Approximate the volume to the nearest unit. In Exercises 2 and 3, use $\pi \approx 3.1416$.

(c) Express the surface area in terms of $\pi$.

1. 2 centimeters
2. 5 inches
3. 10 feet

## SIMILARITY OF SPHERES

*Any two spheres have the same shape, and are therefore similar. Let the radii of the spheres be r and r′, where r > r′. Then the ratio of similarity is given by*

$$\frac{r}{r'},$$

*the ratio of the surface areas is given by*

$$\left(\frac{r}{r'}\right)^2,$$

*and the ratio of the volumes is given by*

$$\left(\frac{r}{r'}\right)^3.$$

EXAMPLE 2. In Figure 2.27, suppose the ratio of similarity of the given spheres is 3.

(a) Find the radius of the smaller sphere.

(b) Find the ratio of the surface areas.

(c) Find the ratio of the volume.

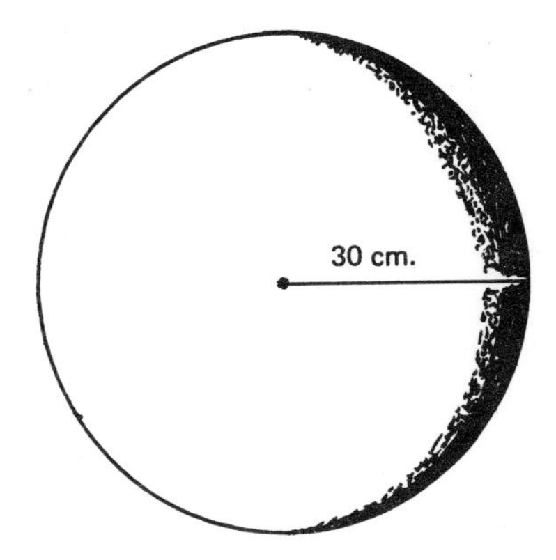


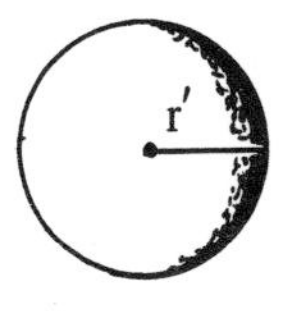


FIGURE 2.27

SOLUTION.

(a)

$$\frac{r}{r'} = 3$$

$$\frac{30 \text{ cm.}}{r'} = 3$$

$$30 \text{ cm.} = 3r'$$

$$10 \text{ cm.} = r'$$

(b) Ratio of surface areas = (Ratio of similarity)$^2$ = $3^2$ = 9

(c) Ratio of volumes = (Ratio of similarity)$^3$ = $3^3$ = 27

## CLASS EXERCISES

4. (a) Find the ratio of similarity of two spheres with radii 40 centimeters and 8 centimeters, respectively.

   (b) Find the ratio of the surface areas.

   (c) Find the ratio of the volumes.

## CYLINDERS

Figure 2.28(a) shows a *(right circular) cylinder.* Its *bases* (both bottom and top) are circular regions each of which has radius r. Its *height* is h. Note that *the radius and the corresponding height form a right angle.*

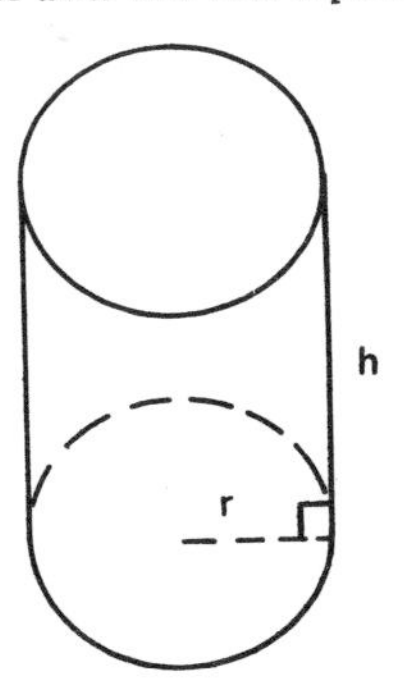


FIGURE 2.28 (a) A cylinder with base radius r and height h.

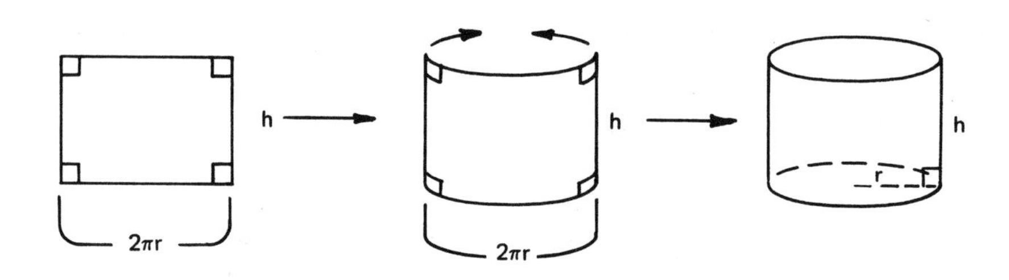


FIGURE 2.28 (b)

*The volume, V, of a cylinder equals the area of a base times the height.* Because the area of a base is $\pi r^2$, it follows that

$$V = \pi r^2 h.$$

A rectangular sheet of paper can be rolled to form the lateral surface of a cylinder of base radius r and height h. (See Figure 2.28 (b).) The lateral surface area is then equal to the area of the sheet. The length of the sheet equals $2\pi r$, the circumference of the circular base, and the width of the sheet equals h, the height of the cylinder. Therefore,

lateral surface area of cylinder = area of sheet

$$= 2\pi r \cdot h$$

Let S be the lateral surface area. Then

$$S = 2\pi rh$$

(If the top and bottom are also included, the *total surface area* would be

| $2\pi rh$ | + | $2\pi r^2$ |
|---|---|---|
| lateral surface area | | $2 \times$ area of circular base (top and bottom) |

EXAMPLE 3. Consider the given cylinder.

(a) Express the volume in terms of $\pi$.

(b) Approximate the volume to the nearest cubic inch. Use $\pi \approx 3.142$.

(c) Express the lateral surface area in terms of $\pi$.

SOLUTION.

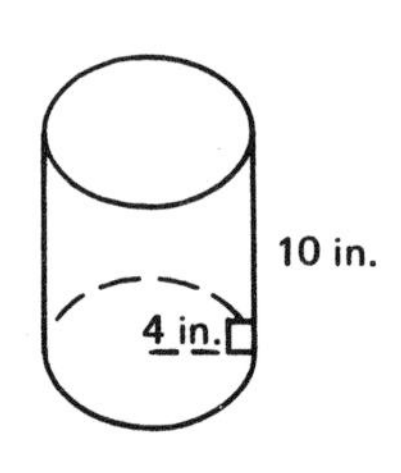


FIGURE 2.29

(a) $V = \pi r^2 h$

$= \pi(4 \text{ in.})^2 \times 10 \text{ in.}$

$= \pi \times 16 \text{ in.}^2 \times 10 \text{ in.}$

$= 160\pi \text{ in.}^3$

(b) $V \approx 160 \times 3.142 \text{ in.}^3$

$= 502.72 \text{ in.}^3$

To the nearest cubic inch,

$V \approx 503 \text{ in.}^3$

(c) $S = 2\pi rh$

$= 2\pi \times 4 \text{ in.} \times 10 \text{ in.}$

$= 80\pi \text{ in.}^2$

EXAMPLE 4. Find the base radius of a cylinder with height 8 centimeters and volume $72\pi$ cubic centimeters.

SOLUTION.

$$V = \pi r^2 h$$

$$72\pi \text{cc.} = \pi r^2 \times 8 \text{ cm.} \qquad \text{cc.} = \text{cm.}^3$$

$$\frac{\overset{9 \text{ cm.}^2}{\cancel{72\pi \text{ cm.}^3}}}{\underset{1}{\cancel{8\pi \text{ cm.}}}} = \frac{\overset{1}{\cancel{8\pi \text{ cm.}}} \times r^2}{\underset{1}{\cancel{8\pi \text{ cm.}}}}$$

$$9 \text{ cm.}^2 = r^2$$

$$3 \text{ cm.} = r$$

## CLASS EXERCISES

In Exercises 5 and 6, consider the given cylinder.

(a) Express the volume in terms of $\pi$.

(b) Approximate the volume to the nearest unit. Use $\pi \approx 3.142$.

(c) Express the lateral surface area in terms of $\pi$.

5.

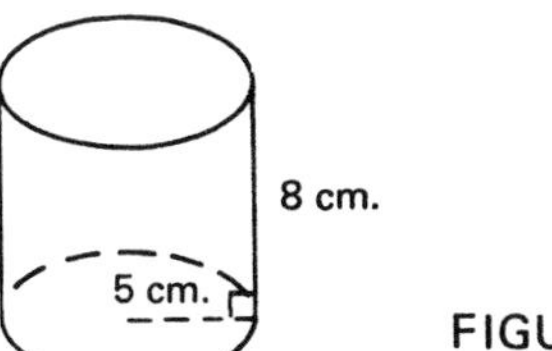


FIGURE 2.30

6.

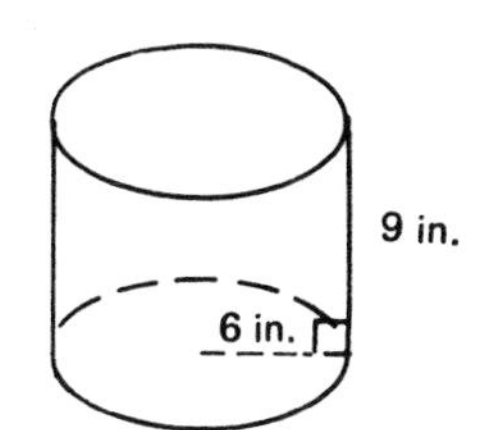


FIGURE 2.31

7. Find the height of a cylinder with base radius 10 inches and volume $400\pi$ cubic inches.

8. Find the base radius of a cylinder with height 5 feet and volume $20\pi$ cubic feet.

### CONES

Figure 2.32 shows a *(right circular) cone.* Its *base* is a circle with radius r. Its *height* is h and its *slant height* is s. The height together with a radius and the corresponding slant height form a right triangle. By the Pythagorean Theorem,

$$s^2 = r^2 + h^2$$

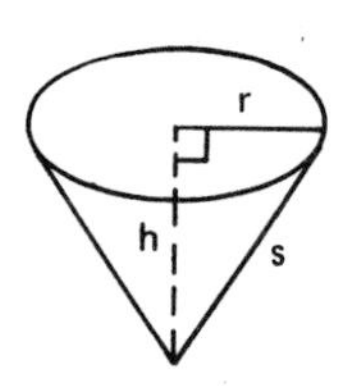


FIGURE 2.32 A cone with base radius r, height h, and slant height s.
$s^2 = r^2 + h^2$

The volume, V, of a cone equals 1/3 the area of the base, $\pi r^2$, times the height, h.

$$V = \frac{1}{3}\pi r^2 h = \frac{\pi r^2 h}{3}$$

The *lateral* surface area, S, is given by

$$S = \pi rs.$$

EXAMPLE 5. Consider a cone with base radius 3 inches and height 4 inches, as in Figure 2.33.

(a) Express the volume in terms of $\pi$.

(b) Approximate the volume to the nearest cubic inch.

(c) Find the slant height.

(d) Express the lateral surface area in terms of $\pi$.

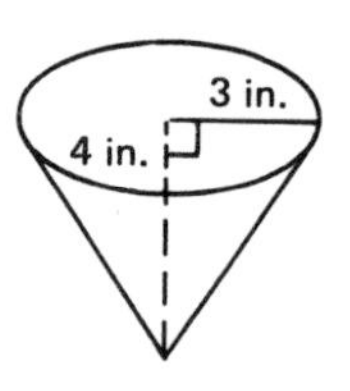


FIGURE 2.33

SOLUTION.

(a) $V = \frac{1}{3}\pi r^2 h$

$= \frac{1}{3}\pi(3 \text{ in.})^2 \times 4 \text{ in.}$

$= \frac{1}{\cancel{3}_1}\pi \, \cancel{9 \text{ in.}^2}^{\,3 \text{ in.}^2} \times 4 \text{ in.}$

$= 12\,\pi \text{ in.}^3$

(b) $V \approx 12 \times 3.14 \text{ in.}^3$

$= 37.68 \text{ in.}^3$

To the nearest cubic inch,

$V \approx 38 \text{ in.}^3$

(c) By the Pythagorean Theorem,

$s^2 = 3^2 \text{ in.}^2 + 4^2 \text{ in.}^2$

$= 9 \text{ in.}^2 + 16 \text{ in.}^2$

$= 25 \text{ in.}^2$

$s = 5 \text{ in.}$

(d) $S = \pi rs$

$= \pi(3 \text{ in.})(5 \text{ in.})$

$= 15\pi \text{ in.}^2$

## CLASS EXERCISES

In Exercises 9 and 10, consider the given cone.

(a) Express the volume in terms of $\pi$.

(b) Approximate the volume to the nearest unit. Use $\pi \approx 3.142$.

(c) Find the slant height.

(d) Express the lateral surface area in terms of $\pi$.

9.

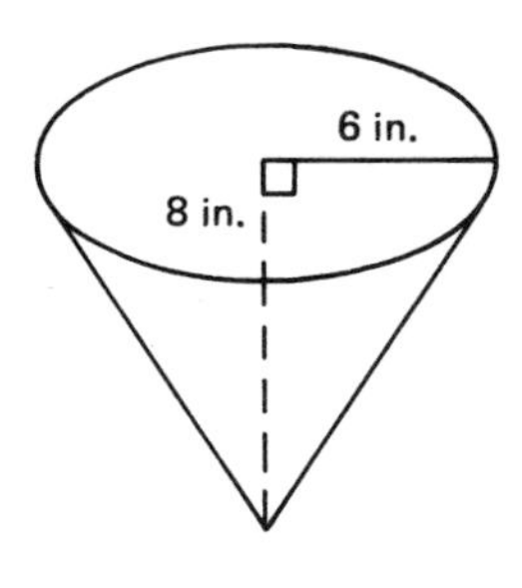


FIGURE 2.34

10.

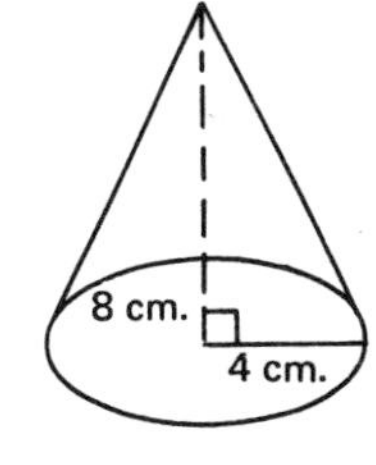


FIGURE 2.35

11. Find the height of a cone with base radius 2 inches and volume $28\pi$ cubic inches.

12. Find the base radius of a cone with height 30 inches and volume $40\pi$ cubic inches.

### SIMILARITY OF CYLINDERS AND OF CONES

Two (right circular) cylinders with base radii r and r′, r > r′, and heights h and h′ are similar if

$$\frac{r}{r'} = \frac{h}{h'}.$$

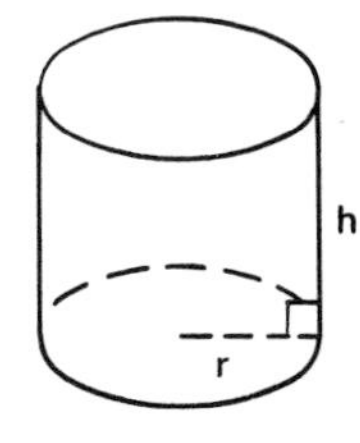


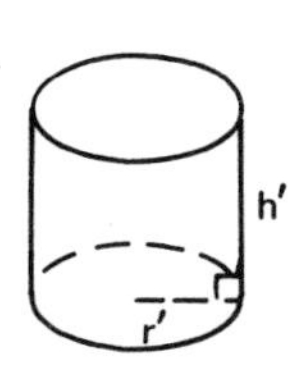


FIGURE 2.36 Similar cylinders.

$$\frac{r}{r'} = \frac{h}{h'}$$

And two (right circular) cones with base radii r and $r'$, $r > r'$, and heights h and $h'$ are similar if

$$\frac{r}{r'} = \frac{h}{h'}.$$

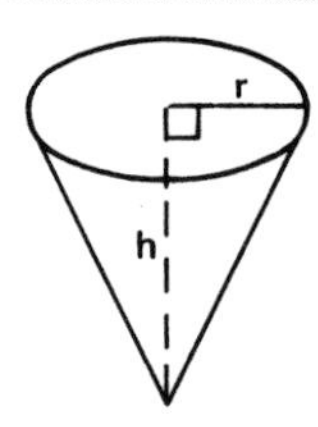

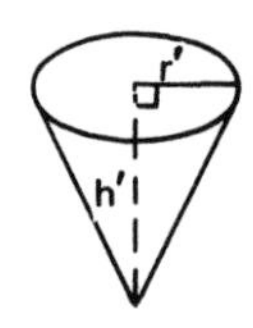


FIGURE 2.37 Similar cones.

$$\frac{r}{r'} = \frac{h}{h'}$$

In each case, the ratio of similarity is $\frac{r}{r'}$, the ratio of the lateral surface areas is $\left(\frac{r}{r'}\right)^2$, and the ratio of the volumes is $\left(\frac{r}{r'}\right)^3$.

EXAMPLE 6. Suppose two cylinders are similar and one has base radius 6 inches and height 9 inches. (See Figure 2.38.)

(a) If the other cylinder has base radius 2 inches, find its height.

(b) What is the ratio of similarity?

(c) What is the ratio of the lateral surface areas?

(d) What is the ratio of the volumes?

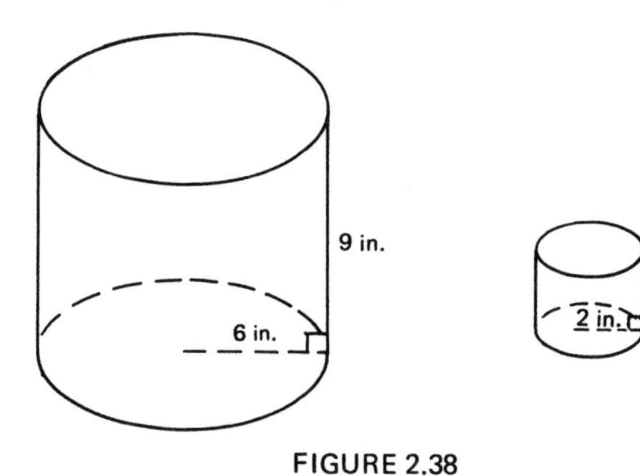


FIGURE 2.38

SOLUTION.

(a)

$$\frac{r}{r'} = \frac{h}{h'}$$

$$\frac{\overset{3}{\cancel{6\text{ in.}}}}{\underset{1}{\cancel{2\text{ in.}}}} = \frac{9\text{ in.}}{h'}$$

$$3h' = 9\text{ in.}$$

$$h' = 3\text{ in.}$$

(b) Ratio of similarity $= \frac{r}{r'} = \frac{6\text{ in.}}{2\text{ in.}} = 3$

(c) Ratio of lateral surface areas = (Ratio of similarity)$^2$ = $3^2$ = 9

(d) Ratio of volumes = (Ratio of similarity)$^3$ = $3^3$ = 27

EXAMPLE 7. Suppose two cones are similar and the ratio of similarity is 2. If the larger cone has base radius 4 inches and height 5 inches, find the volume of the smaller cone. (See Figure 2.39.)

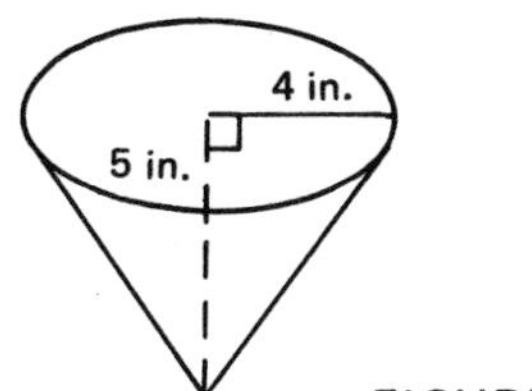


FIGURE 2.39

SOLUTION. Let V be the volume of the larger cone,

r be its base radius,

h be its height,

and let V′ be the volume of the smaller cone.

$$V = \frac{1}{3}\pi r^2 h$$

$$= \frac{1}{3}\pi(4 \text{ in.})^2 \times 5 \text{ in.}$$

$$= \frac{80\pi}{3} \text{ in.}^3$$

Because the ratio of similarity is 2, the ratio of the volumes is $2^3$, or 8.

$$V = 8\,V'$$

$$\frac{80\pi}{3} \text{in.}^3 = 8\,V'$$

$$\frac{80\pi}{3 \times 8} \text{in.}^3 = \frac{8\,V'}{8} \qquad \frac{\frac{80\pi}{3}}{8} = \frac{80\pi}{3 \times 8}$$

$$\frac{10\pi}{3} \text{in.}^3 = V'$$

## CLASS EXERCISES

13. Suppose two cylinders are similar and one has base radius 8 centimeters and height 20 centimeters.

    (a) If the other cylinder has height 5 centimeters, find its base radius.

    (b) What is the ratio of similarity?

(c) What is the ratio of the lateral surface areas?

(d) What is the ratio of the volumes?

14. Suppose the ratio of the volumes of two cones is 27. If the larger cone has height 18 inches, find the height of the smaller cone.

## SOLUTIONS TO CLASS EXERCISES

1. (a) $V = \frac{4}{3}\pi r^3$

$= \frac{4}{3}\pi(2 \text{ cm.})^3$

$= \frac{4}{3}\pi \times 8 \text{ cc.}$ $\qquad$ $\text{cm}^3 = \text{cc.}$ (cubic centimeters)

$= \frac{32\pi}{3} \text{ cc.}$

(b) $V \approx \frac{32 \times 3.14}{3}$

$= \frac{100.48 \text{ cc.}}{3}$

$\approx 33.49 \text{ cc.}$

To the nearest cubic centimeter,

$V \approx 33 \text{ cc.}$

(c) $S = 4\pi r^2$

$= 4\pi (2 \text{ cm.})^2$

$= 4\pi \times 4 \text{ cm.}^2$

$= 16\pi \text{ cm.}^2$

2. (a) $V = \frac{4}{3}\pi r^3$

$= \frac{4}{3}\pi (5 \text{ in.})^3$

$= \frac{4}{3}\pi \times 125 \text{ in.}^3$

$= \frac{500\pi}{3} \text{ in.}^3$

(b) $V \approx \frac{500 \times 3.1416 \text{ in.}^3}{3}$

$= \frac{1570.80}{3} \text{in.}^3$

$= 523.60 \text{ in.}^3$

To the nearest cubic inch,

$V \approx 524 \text{ in.}^3$

(c) $S = 4\pi (5 \text{ in.})^2$

$= 4\pi \times 25 \text{ in.}^2$

$= 100\pi \text{ in.}^2$

3. (a) $V = \frac{4}{3}\pi r^3$

$= \frac{4}{3}\pi (10 \text{ ft.})^3$

$= \frac{4}{3}\pi \times 1000 \text{ ft.}^3$

$= \frac{4000\pi}{3} \text{ft.}^3$

(b) $V = \frac{4000 \times 3.1416}{3} \text{ ft.}^3$

$= \frac{12{,}566.4}{3} \text{ ft.}^3$

$= 4188.8 \text{ ft.}^3$

To the nearest cubic foot,

$V \approx 4189 \text{ ft.}^3$

(c) $S = 4\pi(10 \text{ ft.})^2$

$= 4\pi \times 100 \text{ ft.}^2$

$= 400\pi \text{ ft.}^2$

4. (a) Ratio of similarity $= \frac{r}{r'} = \frac{40 \text{ cm.}}{8 \text{ cm.}} = 5$

(b) Ratio of surface areas $= 5^2 = 25$

(c) Ratio of volumes $= 5^3 = 125$

5. (a) $V = \pi r^2 h$

$= \pi (5 \text{ cm.})^2 \times 8 \text{ cm.}$

$= \pi \times 25 \text{ cm.}^2 \times 8 \text{ cm.}$

$= 200\pi \text{ cc.}$

(b) $V \approx 200 \times 3.142 \text{ cc.} = 628.4 \text{ cc.}$

To the nearest cubic centimeter, $V \approx 628$ cc.

(c) $S = 2\pi rh = 2\pi\ (5\text{ cm.})\ (8\text{ cm.}) = 80\pi\text{ cm.}^2$

6. (a) $V = \pi r^2 h$

$= \pi\ (6\text{ in.})^2 \times 9\text{ in.}$

$= \pi\ 36\text{ in.}^2 \times 9\text{ in.}$

$= 324\pi\text{ in.}^3$

(b) $V \approx 324 \times 3.142\text{ in.}^3$

$= 1018.008\text{ in.}^3$

To the nearest cubic inch,

$V \approx 1018\text{ in.}^3$

(c) $S = 2\pi rh$

$= 2\pi\ (6\text{ in.})\ (9\text{ in.})$

$= 108\pi\text{ in.}^2$

7. $V = \pi r^2 h$

$400\pi\text{ in.}^3 = \pi\ (10\text{ in.})^2 \times h$

$400\pi\text{ in.}^3 = 100\pi\text{ in.}^2 \times h$

$$\frac{400\pi\text{ in.}^3}{100\pi\text{ in.}^2} = \frac{100\pi\text{ in.}^2 \times h}{100\pi\text{ in.}^2}$$

$4\text{ in.} = h$

8. $V = \pi r^2 h$

$20\pi\text{ ft.}^3 = \pi r^2 \times 5\text{ ft.}$

$$\frac{20\pi\text{ ft.}^3}{5\pi\text{ ft.}} = \frac{5\pi\text{ ft.} \times r^2}{5\pi\text{ ft.}}$$

$4\text{ ft.}^2 = r^2$

$2\text{ ft.} = r$

9. (a) $V = \frac{1}{3}\pi r^2 h$

$= \frac{1}{3}\pi\ (6\text{ in.})^2 \times 8\text{ in.}$

$= \frac{1}{\cancel{3}_1}\pi \times \cancel{36\text{ in.}^2}^{\,12\text{ in.}^2} \times 8\text{ in.}$

$= 96\pi\text{ in.}^3$

(b) $V \approx 96 \times 3.142\text{ in.}^3 = 301.632\text{ in.}^3$

To the nearest cubic inch, $V \approx 302\text{ in.}^3$

(c) $s^2 = r^2 + h^2$

$s^2 = (6 \text{ in.})^2 + (8 \text{ in.})^2$

$s^2 = 36 \text{ in.}^2 + 64 \text{ in.}^2$

$s^2 = 100 \text{ in.}^2$

$s = 10 \text{ in.}$

(d) $S = \pi rs$

$= \pi (6 \text{ in.})(10 \text{ in.})$

$= 60\pi \text{ in.}^2$

10. (a) $V = \frac{1}{3}\pi r^2 h$

$= \frac{1}{3}\pi (4 \text{ cm.})^2 \times 8 \text{ cm.}$

$= \frac{1}{3}\pi \times 16 \text{ cm.}^2 \times 8 \text{ cm.}$

$= \frac{128\pi}{3} \text{ cc.}$

(b) $V \approx \frac{128 \times 3.142}{3} \text{ cc.}$

$= \frac{402.176}{3} \text{ cc.}$

$\approx 134.059 \text{ cc.}$

To the nearest cubic centimeter,

$V \approx 134 \text{ cc.}$

(c) $s^2 = r^2 + h^2$

$= (4 \text{ cm.})^2 + (8 \text{ cm.})^2$

$= 16 \text{ cm.}^2 + 64 \text{ cm.}^2$

$= 80 \text{ cm.}^2$

$s = \sqrt{80} \text{ cm.}$

$= \sqrt{16 \cdot 5} \text{ cm.}$

$= \sqrt{16}\ \sqrt{5} \text{ cm.}$

$= 4\sqrt{5} \text{ cm.}$

(d) $S = \pi rs$

$= \pi\,(4\text{ cm.})\,(4\sqrt{5}\text{ cm.})$

$= 16\sqrt{5}\pi\text{ cm.}$

11. $V = \frac{1}{3}\pi r^2 h$

$28\pi\text{ in.}^3 = \frac{1}{3}\pi\,(2\text{ in.}^2) \times h$

$84\pi\text{ in.}^3 = 4\pi\text{ in.}^2 \times h$

$\frac{84\pi\text{ in.}^3}{4\pi\text{ in.}^2} = \frac{4\pi\text{ in.}^2 \times h}{4\pi\text{ in.}^2}$

$21\text{ in.} = h$

12. $V = \frac{1}{3}\pi r^2 h$

$40\pi\text{ in.}^3 = \frac{1}{\cancel{3}_{\,1}}\pi r^2 \times \overset{10\text{ in.}}{\cancel{30\text{ in.}}}$

$\frac{40\pi\text{ in.}^3}{10\pi\text{ in.}} = \frac{10\pi\text{ in.} \times r^2}{10\pi\text{ in.}}$

$4\text{ in.}^2 = r^2$

$2\text{ in.} = r$

13. (a) $\frac{r}{r'} = \frac{h}{h'}$

$\frac{8\text{ cm.}}{r'} = \frac{\overset{4}{\cancel{20\text{ cm.}}}}{\underset{1}{\cancel{5\text{ cm.}}}}$

$8\text{ cm.} = 4r'$

$2\text{ cm.} = r'$

(b) Ratio of similarity $= \frac{r}{r'} = \frac{8\text{ cm.}}{2\text{ cm.}} = 4$

(c) Ratio of lateral surface areas $= 4^2 = 16$

(d) Ratio of volumes $= 4^3 = 64$

14. Ratio of volumes = (Ratio of similarity)$^3$

$27 = (\text{ratio of similarity})^3$ Find the cube root of each side.

$3 = \text{Ratio of similarity}$

$$\frac{h}{h'} = 3$$

$$\frac{18 \text{ in.}}{h'} = 3$$

$$18 \text{ in.} = 3h'$$

$$6 \text{ in.} = h'$$

NAME

CLASS

ANSWERS

## HOME EXERCISES

In Exercises 1-3, consider a sphere with the given radius.

(a) Express the volume in terms of $\pi$.

(b) Approximate the volume to the nearest unit, using $\pi \approx 3.142$.

(c) Express the surface area in terms of $\pi$.

1. 4 kilometers
2. 6 centimeters
3. 9 meters

In the remaining approximations, unless otherwise specified, use $\pi \approx 3.14$.

In Exercises 4 and 5, consider a sphere with the given radius.

(a) Express the volume in terms of $\pi$.

(b) Approximate the volume to the nearest tenth of an inch.

(c) Express the surface area in terms of $\pi$.

4. .3 inch
5. ½ inch

6. Find the radius of a sphere whose volume is $\frac{4\pi}{3}$ cubic centimeters.

7. Find the radius of a sphere whose surface area is $16\pi$ square yards.

8. Suppose the ratio of similarity of two spheres is 3, and the larger sphere has radius 6 centimeters.

   (a) Find the radius of the smaller sphere.

   (b) Find the ratio of the surface areas.

   (c) Find the ratio of the volumes.

1. (a) ______ (b) ______ (c) ______
2. (a) ______ (b) ______ (c) ______
3. (a) ______ (b) ______ (c) ______
4. (a) ______ (b) ______ (c) ______
5. (a) ______ (b) ______ (c) ______
6. ______
7. ______
8. (a) ______ (b) ______ (c) ______

ANSWERS

9. ____________

10. ____________

11. (a) ____________

(b) ____________

(c) ____________

12. (a) ____________

(b) ____________

(c) ____________

13. (a) ____________

(b) ____________

(c) ____________

14. (a) ____________

(b) ____________

(c) ____________

9. Suppose the ratio of similarity of two spheres is 2, and the smaller sphere has radius 2 millimeters. Find the volume of the larger sphere.

10. Suppose the volumes of two spheres are $40\pi$ cubic inches and $5\pi$ cubic inches. What is the ratio of similarity?

In Exercises 11-14, consider the given cylinder.

(a) Express the volume in terms of $\pi$.

(b) Approximate the volume to the nearest unit, using $\pi \approx 3.142$.

(c) Express the lateral surface area in terms of $\pi$.

11. base radius 4 inches, height 10 inches

12. base radius 5 centimeters, height 5 centimeters

13.

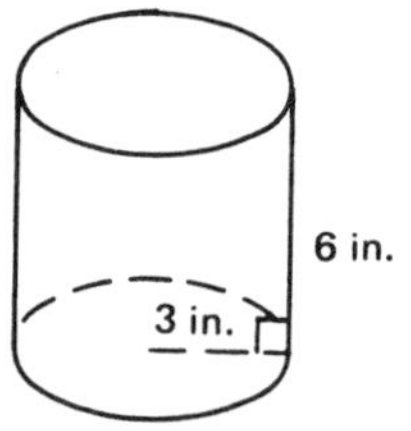

FIGURE 2.40

14.

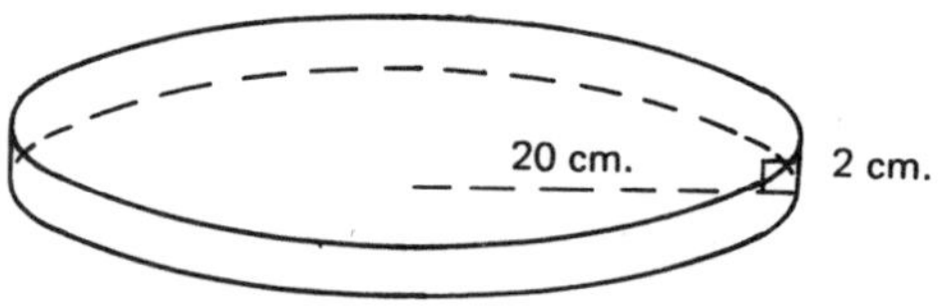

FIGURE 2.41

NAME ____________

CLASS ____________

ANSWERS

In Exercises 15 and 16, find the height of the given cylinder.

15. base radius 5 inches, volume $150\pi$ cubic inches

16. base radius 6 centimeters, volume $72\pi$ cubic centimeters

In Exercises 17 and 18, find the base radius of the given cylinder.

17. height 4 inches, volume $400\pi$ cubic inches

18. height 5 centimeters, volume $20\pi$ cubic centimeters

In Exercises 19-22, consider the given cone.

(a) Express the volume in terms of $\pi$.

(b) Approximate the volume to the nearest unit.

(c) Express the lateral surface area in terms of $\pi$.

19. base radius 3 inches, height 10 inches

20. base radius 7 centimeters, height 6 centimeters

21.

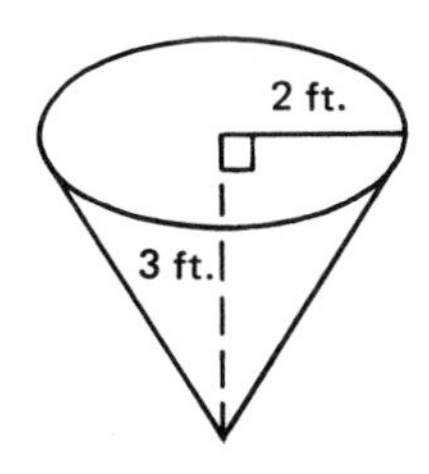


FIGURE 2.42

22.

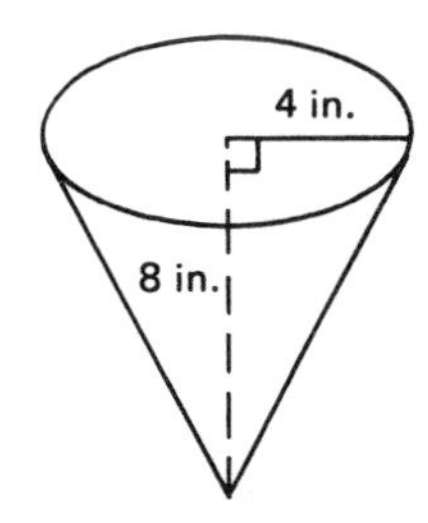


FIGURE 2.43

15. ____________

16. ____________

17. ____________

18. ____________

19. (a) ____________

(b) ____________

(c) ____________

20. (a) ____________

(b) ____________

(c) ____________

21. (a) ____________

(b) ____________

(c) ____________

22. (a) ____________

(b) ____________

(c) ____________

ANSWERS

23. ____________

24. ____________

25. ____________

26. ____________

27. ____________

28. ____________

29. ____________

30. ____________

31. ____________

In Exercises 23 and 24, find the height of the given cone.

23. base radius 3 inches, volume $18\pi$ cubic inches

24. base radius 2 centimeters, volume $4\pi$ cubic centimeters

In Exercises 25 and 26, find the base radius of the given cone.

25. height 5 inches, volume $15\pi$ cubic inches

26. height 2 centimeters, volume $18\pi$ cubic centimeters

27. Suppose the volume of the cylinder in Figure 2.44 is $300\pi$ cubic inches. Find the volume of the cone.

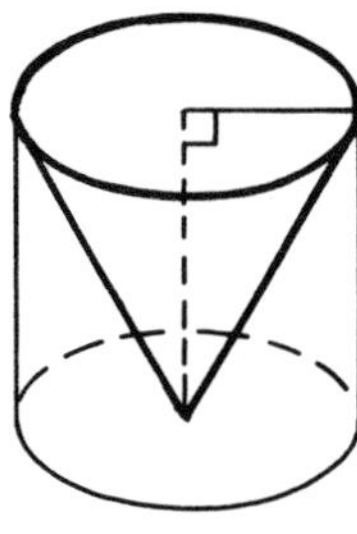

FIGURE 2.44

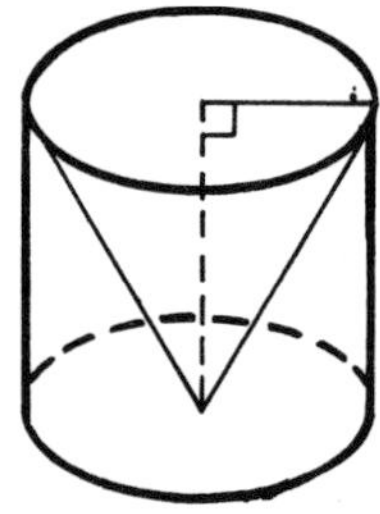

FIGURE 2.45

28. Suppose the volume of the cone in Figure 2.45 is $100\pi$ cubic centimeters. Find the volume of the cylinder.

29. A volleyball has a 5-inch radius. Find its volume.

30. A cylindrical tin can has base radius 2 inches and height 5 inches. Find its volume.

31. A conical sand pile has base radius 4 inches and height 7 inches. (See Figure 2.46.) Find its volume.

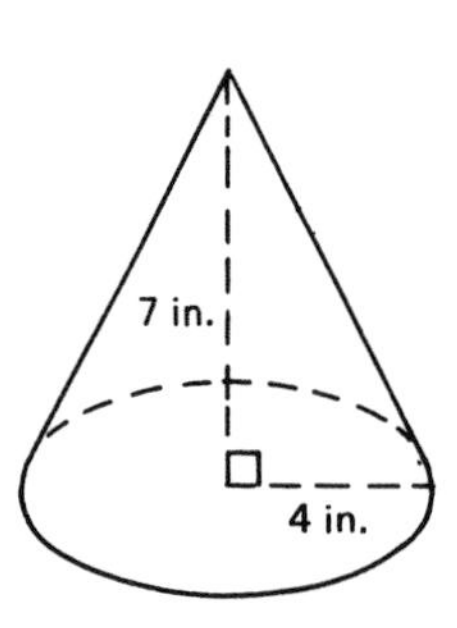


FIGURE 2.46
A conical sand pile.

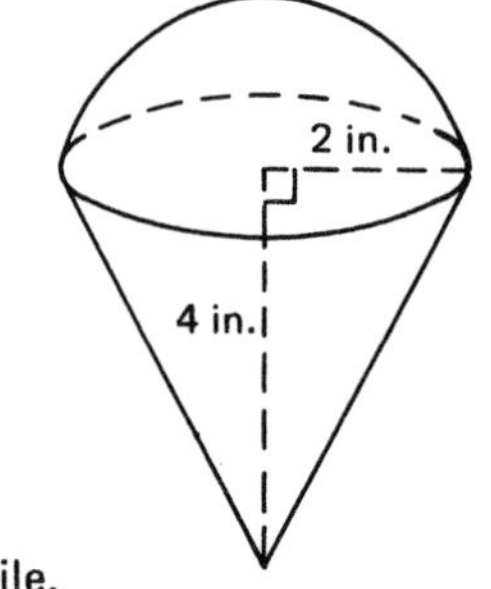


FIGURE 2.47
Ice cream cone.

NAME

CLASS

ANSWERS

32. A cone is filled with ice cream and is topped with a *hemisphere* (half of a sphere) of ice cream. (See Figure 2.47.) The radius of the hemisphere and of the base of the cone is 2 inches. The height of the cone is 4 inches. Find the volume of ice cream.

33. Suppose two cylinders are similar and one has base radius 4 inches and height 8 inches.

    (a) If the other cylinder has base radius 1 inch, find its height.

    (b) What is the ratio of similarity?

    (c) What is the ratio of the lateral surface areas?

    (d) What is the ratio of the volumes?

In Exercises 34 and 35, find the ratio of similarity of the given cylinders.

34.

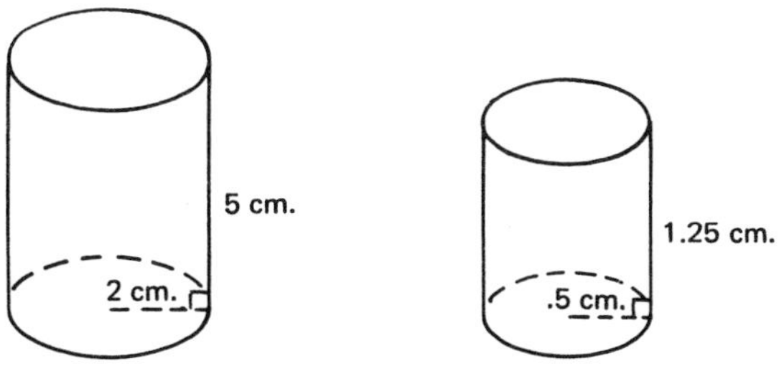

5 cm.

2 cm.

1.25 cm.

.5 cm.

FIGURE 2.48

35.

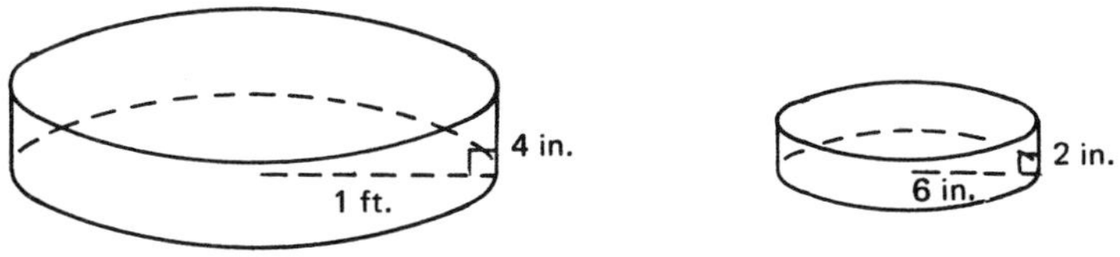

4 in.

1 ft.

2 in.

6 in.

FIGURE 2.49

36. A cylindrical gas storage tank has base radius 30 feet and height 40 feet. A similar tank has height 20 feet. What volume of gas can the smaller tank store?

37. Suppose two cones are similar and the ratio of similarity is 3. If the smaller cone has base radius 5 inches, find the base radius of the larger cone.

32. ______

33. (a) ______

(b) ______

(c) ______

(d) ______

34. ______

35. ______

36. ______

37. ______

ANSWERS

38. ____________________

39. ____________________

38. Suppose two cones are similar and the ratio of similarity is 4. If the smaller cone has base radius 2 inches and height 3 inches, find the volume of the larger cone.

39. Suppose two cones are similar and the ratio of their volumes is 125. If the larger cone has height 10 inches, find the height of the smaller cone.

*Let's Review Chapter 2.*

2.1 Volume and Surface Area

1. Find (a) the volume and (b) the surface area of a rectangular box with length 4 inches, width 3 inches, and height 5 inches.

2. Find (a) the volume and (b) the surface area of a cube of side length 4 centimeters.

3. Suppose the rectangular boxes in Figure 2.50 are similar.

   (a) Find w.

   (b) What is the ratio of similarity?

   (c) Find h′.

   (d) What is the ratio of the surface areas?

   (e) What is the ratio of the volumes?

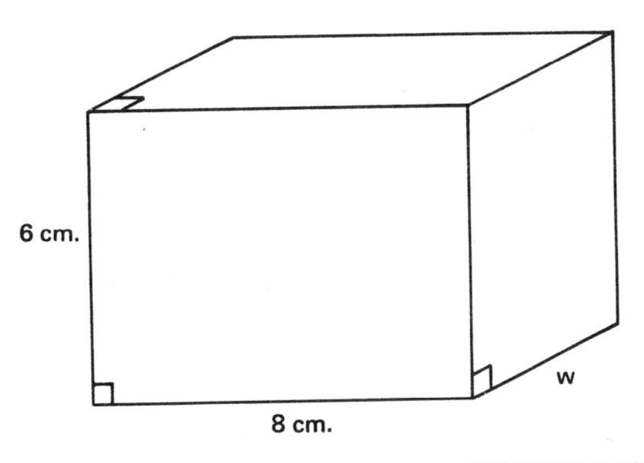


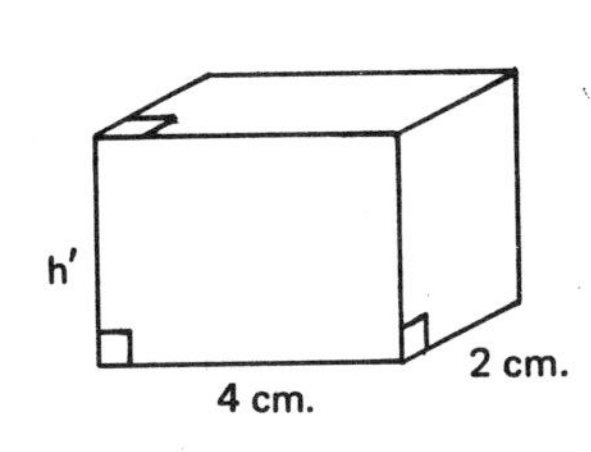


FIGURE 2.50

2.2 Spheres, Cylinders, Cones

4. Consider a sphere of radius 4 inches.

   (a) Express the volume in terms of $\pi$.

(b) Approximate the volume to the nearest cubic inch.

(c) Express the surface area in terms of $\pi$.

5. Suppose the ratio of similarity of two spheres is 3, and the larger sphere has radius 9 centimeters. Find the volume of the smaller sphere.

6. Consider a cylinder with base radius 3 inches and height 4 inches.

   (a) Express the volume in terms of $\pi$.

   (b) Approximate the volume to the nearest cubic inch.

   (c) Express the lateral surface area in terms of $\pi$.

7. Find the base radius of a cylinder with height 8 centimeters and volume $32\pi$ cubic centimeters.

8. Consider a cone with base radius 4 inches and height 6 inches.

   (a) Express the volume in terms of $\pi$.

   (b) Approximate the volume to the nearest cubic inch.

   (c) Express the lateral surface area in terms of $\pi$.

*Try These Exam Questions for Practice.*

1. Find (a) the volume and (b) the surface area of a rectangular box with length 9 inches, width 5 inches, and height 10 inches.

2. A cube has volume 1000 cubic centimeters. Find the length of a side.

3. Find (a) the volume and (b) the surface area of a sphere with radius 10 inches. Express your answers in terms of $\pi$.

4. Suppose two cylinders are similar with ratio of similarity 2. If the smaller cylinder has base radius 2 inches and height 5 inches, find the volume of the larger cylinder.

5. To the nearest cubic centimeter, express the volume of a cone with base radius 6 centimeters and height 8 centimeters. (Use $\pi \approx 3.14$.)

# Chapter 3
# ANGLES

## DIAGNOSTIC TEST FOR CHAPTER 3

Perhaps you are already familiar with some of the material in Chapter 3. This test will indicate which sections in Chapter 3 you need to study. The question number refers to the corresponding section in Chapter 3. If you answer *all* parts of the question correctly, you may omit the section. But *if any part of your answer is wrong, you should study the section.* When you have completed the test, turn to page 168 for the answers.

3.1 Intersecting Lines

(a) Convert $25.4°$ into degrees and minutes.

(b) Find the measure of the complement of a $36°22'$-angle.

(c) In Figure 3A, find the measure of $\angle ABC$.

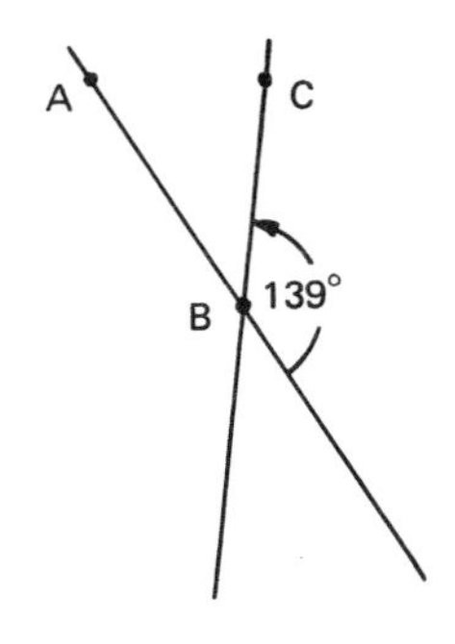


FIGURE 3A

3.2 More on Triangles

(a) Each of the base angles of an isosceles triangle measures $41°$. Find the measure of the third angle.

(b) Find the area of a $30°, 60°, 90°$-triangle whose hypotenuse is 6 centimeters.

(c) Find the side length of an equilateral triangle whose area is $25\sqrt{3}$ inches.

3.3 More on Circles

Refer to Figure 3B. O is the center of the circle. Find:

(a) the measure of $\angle ABC$ (b) the length of $\widehat{AC}$ (c) the area of sector OCA

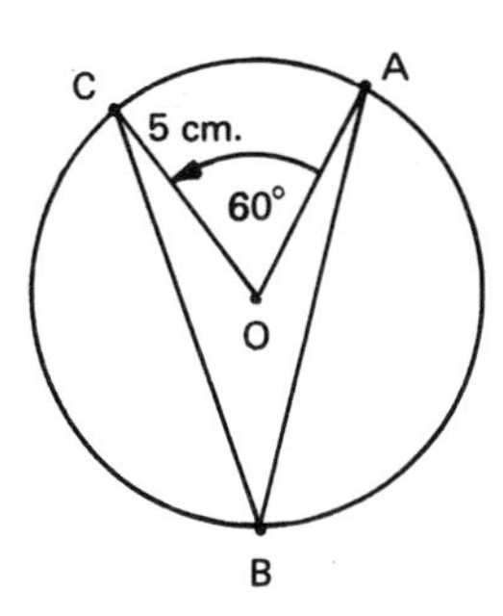


FIGURE 3B

## 3.1 Intersecting Lines

### IDENTIFYING ANGLES

Recall that angles are formed when lines intersect. The intersecting lines are called the *sides* of the angle, and the point of intersection is called the *vertex.* To *identify* an angle, we often use points on their sides. Thus, in Figure 3.1,

$\angle 1 = \angle BAC$ (Read: Angle 1 = Angle BAC)

$\angle 2 = \angle CAD$

Note that A, the *second* letter in each case, represents the vertex. We often speak of the line segments AB and AC as the sides of $\angle 1$ (Recall that line segment AB is the part of the line between points A and B.) In Figure 3.2, there is only one angle indicated, and we call it $\angle$ A. The shaded region is called the *interior of* $\angle$A.

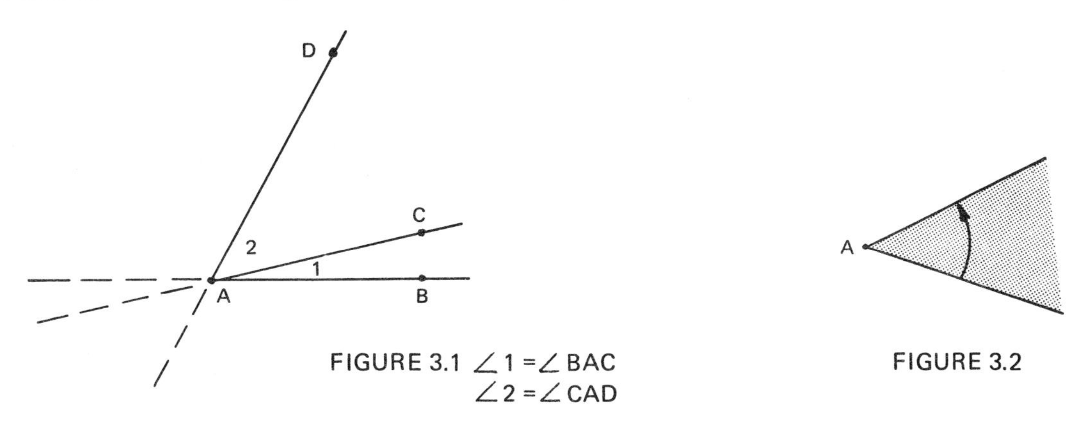


FIGURE 3.1 $\angle 1 = \angle BAC$
$\angle 2 = \angle CAD$

FIGURE 3.2

### MEASURING ANGLES

An angle is *measured* in *degrees.* Write

45° for 45 degrees.

Examples of angles and their measurements are shown in Figure 3.3.

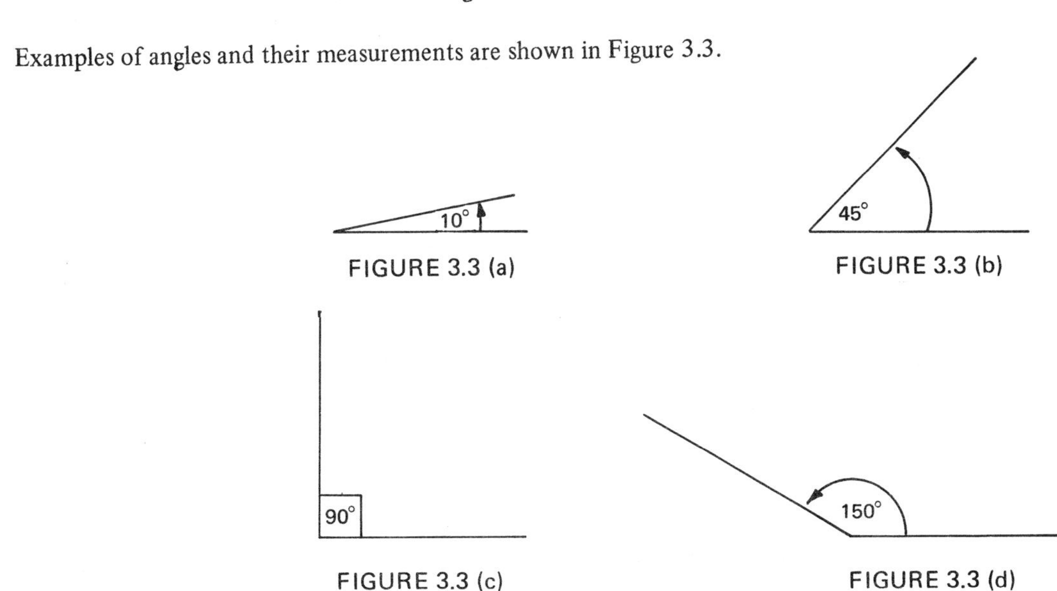


FIGURE 3.3 (a)

FIGURE 3.3 (b)

FIGURE 3.3 (c)

FIGURE 3.3 (d)

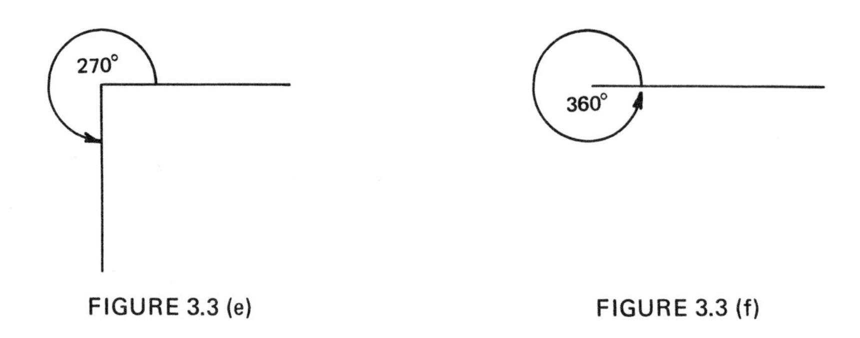


FIGURE 3.3 (e)

FIGURE 3.3 (f)

Usually, an angle measures between 0° and 360°. There are times when it is convenient to consider angles of more than 360°. For example, if a moth circles around a light bulb twice, you might say it flies 2 x 360° or 720°. See Figure 3.4.

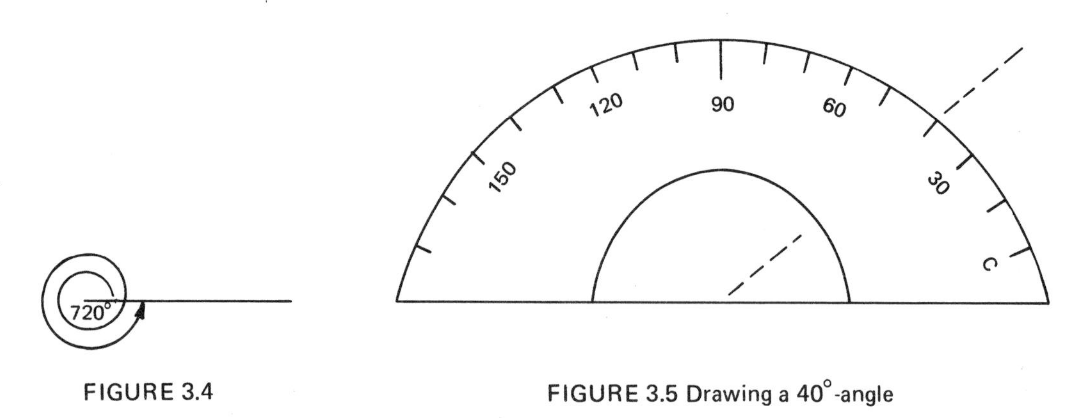


FIGURE 3.4

FIGURE 3.5 Drawing a 40°-angle

To draw an angle of, say 40°, a protractor can be used. See Figure 3.5.

EXAMPLE 1. Use a protractor to draw an angle of (a) 20° (b) 135°.

SOLUTION.

(a) (b)

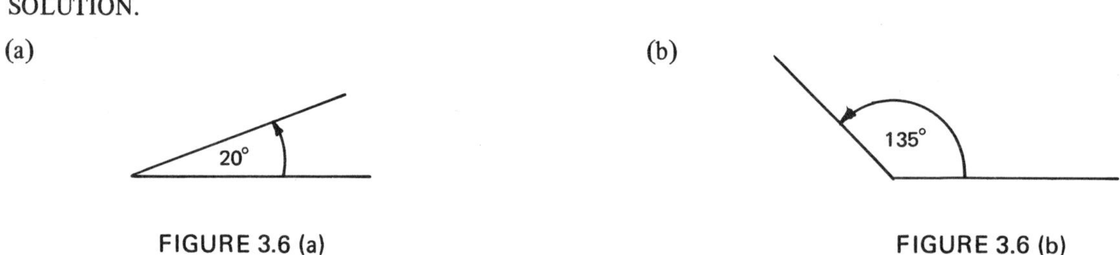


FIGURE 3.6 (a)

FIGURE 3.6 (b)

Finer measurements of angles employ degrees and *minutes.* Each degree equals 60 minutes (60′). Alternatively, decimals can be used. Thus, 20.3° means 20 and three-tenths degrees, whereas 20°3′ means 20 degrees and 3 minutes.

To convert from decimal notation to degrees and minutes, observe that

$$.1° = .1 \times 60 \text{ minutes}$$

Thus,

$$.1° = 6 \text{ minutes}$$

EXAMPLE 2. Convert into degrees and minutes: (a) $.4^\circ$ (b) $12.4^\circ$

SOLUTION. (a) $.4^\circ = 4 \times .1^\circ$

$= 4 \times 6$ minutes

$= 24$ minutes

(b) $12.4^\circ = 12^\circ + .4^\circ$

$= 12^\circ 24'$

EXAMPLE 3. Express $17^\circ 33'$ in decimal notation.

SOLUTION. 60 minutes = 1 degree

First, express the fraction $\frac{33}{60}$ as a decimal.

$$6.0 \overline{)3.30}^{\;.55}$$

Thus, $17^\circ 33' = 17.55^\circ$

Write $\angle A = 40^\circ$ when $\angle A$ *measures* $40^\circ$.

## CLASS EXERCISES

1. Convert $32.7^\circ$ into degrees and minutes.
2. Convert $56^\circ 27'$ into decimal notation.

## CLASSIFYING ANGLES

As you know, a *right angle* measures $90^\circ$. The lines forming a right angle are said to be *perpendicular*. Write

$$L_1 \perp L_2$$

when lines $L_1$ and $L_2$ are perpendicular. An angle that measures $180^\circ$ is called a *straight angle*. An *acute angle* is an angle that measures between $0^\circ$ and $90^\circ$, whereas an *obtuse angle* is one that measures between $90^\circ$ and $180^\circ$. Thus,

$\angle$ **A** *is acute if* $0^\circ < \angle A < 90^\circ$;

$\angle$ **B** *is obtuse if* $90^\circ < \angle B < 180^\circ$.

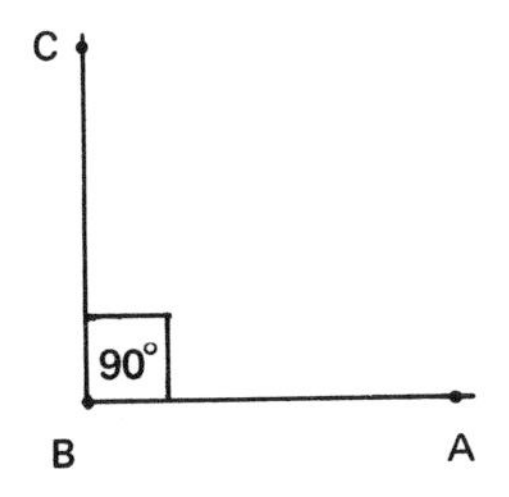


FIGURE 3.7 (a)
A right angle measures $90^\circ$.
$AB \perp BC$

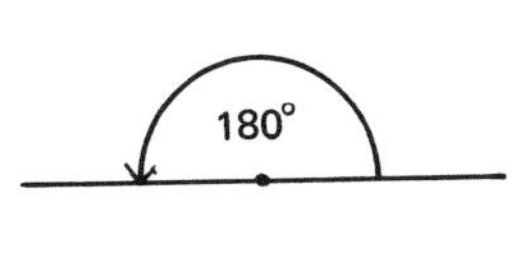


FIGURE 3.7 (b)
A straight angle measures $180^\circ$.

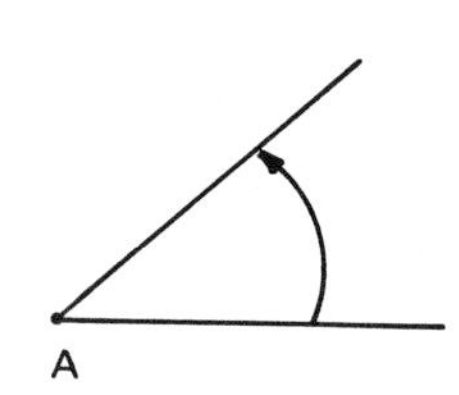


FIGURE 3.7 (c)
$\angle A$ is acute because $0° < \angle A < 90°$.

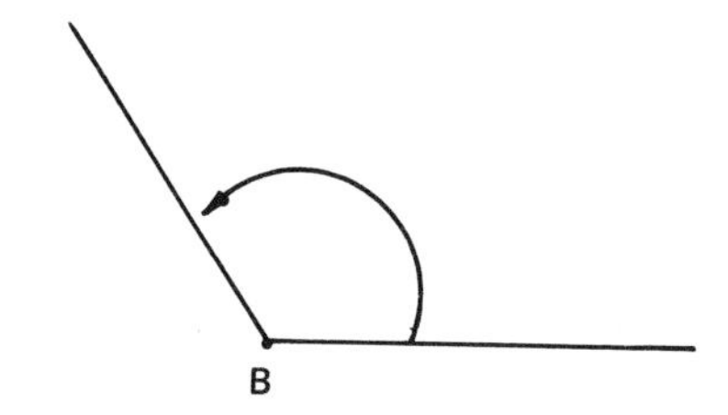


FIGURE 3.7 (d)
$\angle B$ is obtuse because $90° < \angle B < 180°$.

EXAMPLE 4. Classify each angle as either a right angle, a straight angle, an acute angle, or an obtuse angle.

(a) 72° (b) 172° (c) 180°

SOLUTION. (a) acute (b) obtuse (c) straight

Write $\angle 3 = \angle 1 + \angle 2$

if (the measure of) $\angle 3$ is the sum of (the measures of) angles 1 and 2. In this case, $\angle 3$ is called the *sum of angles* 1 *and* 2. If $\angle 1 + \angle 2 = 90°$, each of these angles is the *complement* of the other. Thus, the complement of a 60°-angle is a 30°-angle. If $\angle 1 + \angle 2 = 180°$, each of these angles is the *supplement* of the other. The supplement of a 130°-angle is a 50°-angle.

EXAMPLE 5. Suppose $\angle A = 53°20'$. Find the measure of (a) its complement, (b) its supplement.

SOLUTION. (a) To find the measure of the complement of $\angle A$, consider:

$$\begin{array}{r} 90° \\ \underline{-53°20'} \end{array}$$

In order to subtract *minutes,* borrow 60 minutes (1 degree) from 90°.

$$\begin{array}{r} 89°60' \\ \not{90°} \\ \underline{-53°20'} \\ 36°40' \end{array}$$

The complement of $\angle A$ measures 36°40′.

(b)

$$\begin{array}{r} 179°60' \\ \not{180°} \\ \underline{-\ 53°20'} \\ 126°40' \end{array}$$

The supplement of $\angle A$ measures 126°40′.

Note that *measure of supplement* of $\angle A = 90° +$ *measure of complement* of $\angle A$.

EXAMPLE 6. The complement of $\angle A$ measures three times as much as $\angle A$. Find the measure of $\angle A$.

SOLUTION. Let x be the measure of $\angle A$. Reword the first sentence.

The measure of the complement of ∠ A is three times the measure of ∠ A.

$$90 - x = 3x$$

$$90 = 4x$$

$$\frac{90}{4} = x$$

$$22.5 = x$$

$\angle A = 22.5°$, or $22°30'$

Angles that share a common side are known as *adjacent angles.* In Figure 3.8, ∠ 1 and ∠ 2 share a common side, AC. In Figure 3.9, ∠ 1 and ∠ 2 are *equal* adjacent angles. Their common side, AC, is called the *bisector of* ∠ BAD.

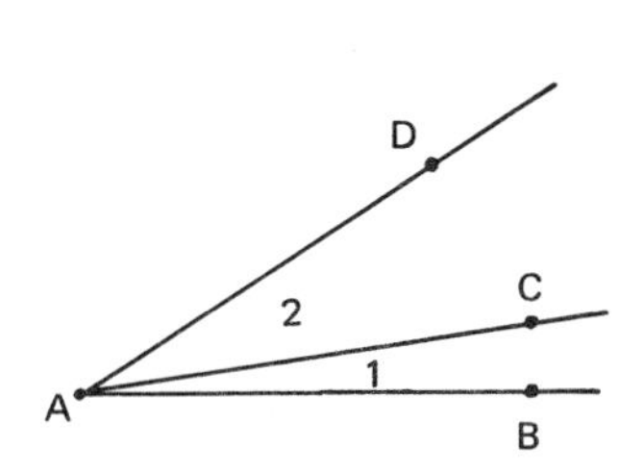


FIGURE 3.8
∠1 and ∠2 are adjacent angles.
They share the common side AC.
∠1 + ∠2 = ∠BAD

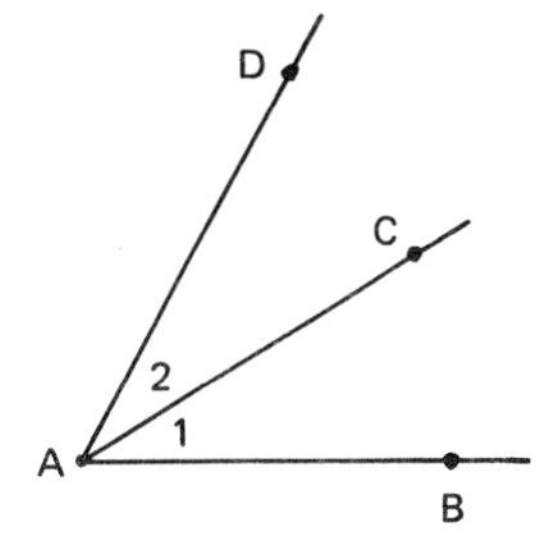


FIGURE 3.9
∠1 = ∠2 and ∠1 + ∠2 = ∠BAD.
Side AC bisects ∠BAD.

When lines intersect, 4 angles are formed. In Figure 3.10, ∠ 1 and ∠ 3 are called *vertical angles,* as are ∠ 2 and ∠ 4. Note that

$$\angle 1 + \angle 2 = 180°$$

$$\angle 3 + \angle 2 = 180°$$

From the first equation,

$$\angle 1 = 180° - \angle 2$$

And from the second equation,

$$\angle 3 = 180° - \angle 2$$

Thus,

$$\angle 1 = \angle 3$$

In general, *vertical angles are equal (in measure).*

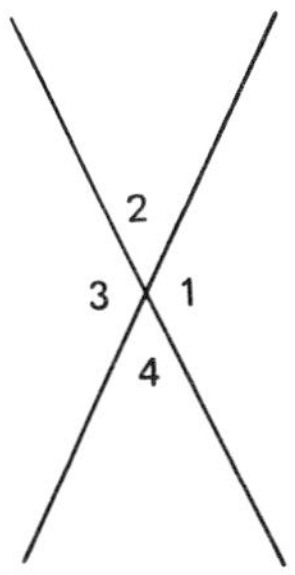

FIGURE 3.10
∠1 and ∠3 are vertical angles;
∠2 and ∠4 are vertical angles. ∠1 = ∠3; ∠2 = ∠4

## CLASS EXERCISES

In Exercises 3-5, suppose $\angle A = 85°30'$.

3. Is $\angle A$ acute or obtuse?

4. Find the complement of $\angle A$.

5. Find the supplement of $\angle A$.

6. The supplement of $\angle B$ measures four times as much as the complement of $\angle B$. Find the measure of $\angle B$.

## TRANSVERSALS

A line that intersects two or more other lines is known as the *transversal* of those lines. In Figure 3.11, L is the transversal of lines $\ell_1$ and $\ell_2$. The pairs of angles,

$\angle 1$ and $\angle 5$, $\angle 2$ and $\angle 6$,

$\angle 3$ and $\angle 7$, $\angle 4$ and $\angle 8$,

are called *corresponding angles.* The pairs of angles

$\angle 1$ and $\angle 7$, $\angle 2$ and $\angle 8$,

are called *alternate interior angles,* whereas the pairs of angles,

$\angle 3$ and $\angle 5$, $\angle 4$ and $\angle 6$,

are known as *alternate exterior angles.*

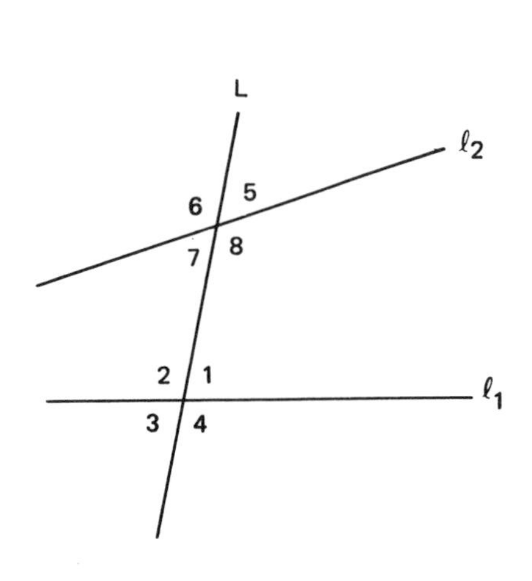


FIGURE 3.11

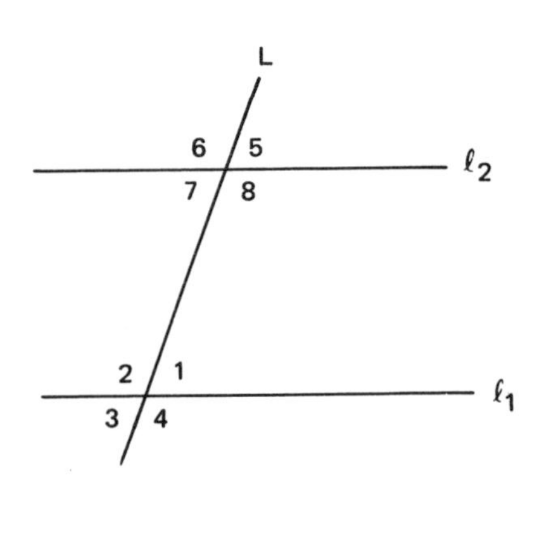


FIGURE 3.12 $\ell_1 \parallel \ell_2$
$\angle 1 = \angle 5, \angle 2 = \angle 8, \angle 4 = \angle 6$

Recall that *parallel lines* are lines in a plane that do not intersect. Write

$$\ell_1 \parallel \ell_2$$

when lines $\ell_1$ and $\ell_2$ are parallel. Consider a *transversal of parallel lines*, as in Figure 3.12. Then

*corresponding angles are equal (in measure),*

*alternate interior angles are equal,*

*and*

*alternate exterior angles are equal.*

In Figure 3.12, $\ell_1 \parallel \ell_2$. Therefore,

$\angle 1 = \angle 5$ (Corresponding angles are equal.)

$\angle 2 = \angle 8$ (Alternate interior angles are equal.)

$\angle 4 = \angle 6$ (Alternate exterior angles are equal.)

Now suppose that two arbitrary lines, $\ell_1$ and $\ell_2$, are cut by a transversal.

*If corresponding angles are equal, then $\ell_1 \parallel \ell_2$.*

Similarly,

*if alternate exterior angles are equal, then $\ell_1 \parallel \ell_2$.*

And

*if alternate exterior angles are equal, then $\ell_1 \parallel \ell_2$.*

EXAMPLE 7. In Figure 3.13, suppose that $\angle 4 = 100°$ and $\angle 5 = 80°$. Show that $\ell_1 \parallel \ell_2$.

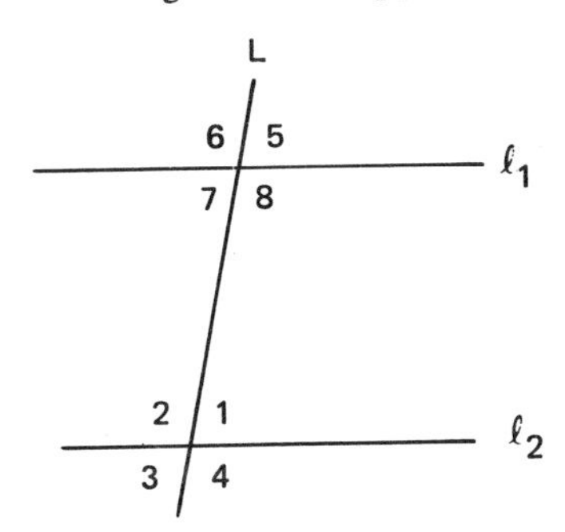


FIGURE 3.13 $\angle 4 = 100°$, $\angle 5 = 80°$

SOLUTION. $\angle 1$ = supplement $\angle 4$

$= 180° - 100°$

$\angle 1 = 80°$

Thus,

$\angle 1 = \angle 5 = 80°$

$\ell_1 \parallel \ell_2$ because corresponding angles are equal. (What other angles could you have used to show that $\ell_1 \parallel \ell_2$ ?)

Write

$$AB = 7 \text{ inches}$$

if the length of the line segment AB is 7 inches.

When three (or more) parallel lines are cut by two (or more) transversals, the lengths of corresponding line segments of the transversals are proportional. In Figure 3.14, $\ell_1$, $\ell_2$, and $\ell_3$ are parallel. (Write $\ell_1 \parallel \ell_2 \parallel \ell_3$.)

$$\frac{AB}{BC} = \frac{DE}{EF}$$

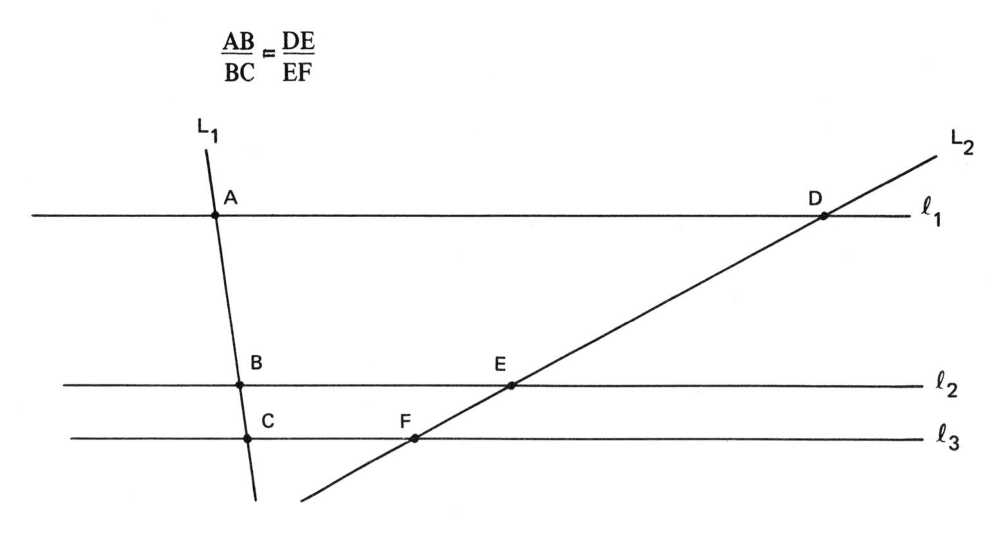


FIGURE 3.14 $\ell_1 \parallel \ell_2 \parallel \ell_3$

$$\frac{AB}{BC} = \frac{DE}{EF}$$

EXAMPLE 8. In Figure 3.14, $\ell_1 \parallel \ell_2 \parallel \ell_3$. Suppose AB = 5 in., BC = 2 in., DE = 15 in. Find the length of EF.

SOLUTION.

$$\frac{AB}{BC} = \frac{DE}{EF}$$

$$\frac{5 \text{ in.}}{2 \text{ in.}} = \frac{15 \text{ in.}}{x \text{ in.}}$$

Cross-multiply:

$$5x = 2 \cdot 15$$

$$5x = 30$$

$$x = 6$$

Thus, EF = 6 inches

## CLASS EXERCISES

In Exercises 7-13, refer to Figure 3.15. Suppose $\angle 1 = 75°$. Find the measure of each angle.

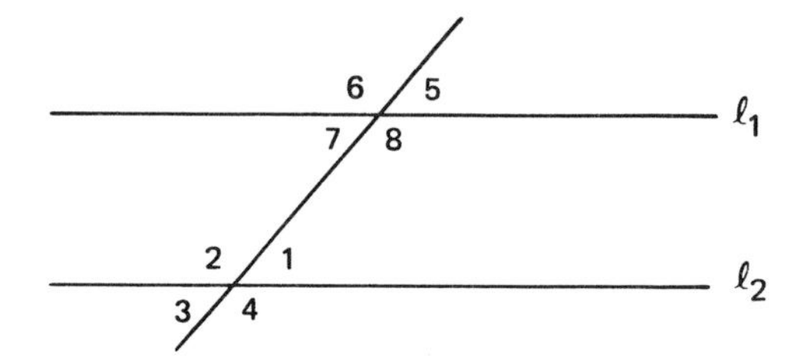


FIGURE 3.15 $\ell_1 \parallel \ell_2$

7. $\angle 2$    8. $\angle 3$    9. $\angle 4$    10. $\angle 5$

11. $\angle 6$    12. $\angle 7$    13. $\angle 8$

14. In Figure 3.16, $\ell_1 \parallel \ell_2 \parallel \ell_3$. Suppose AB = 4 inches, DE = 12 inches, EF = 15 inches. Find the length of BC.

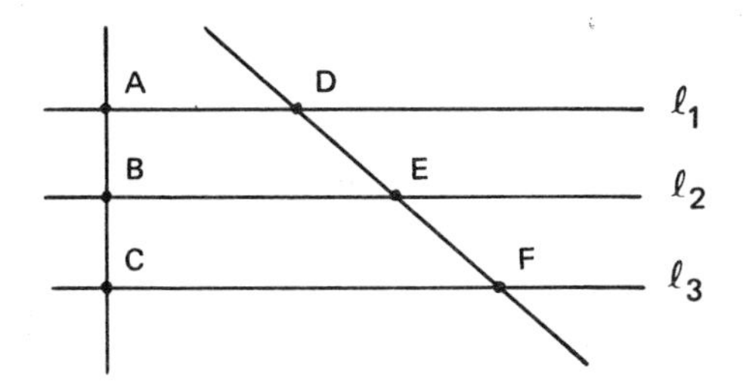


FIGURE 3.16 $\ell_1 \parallel \ell_2 \parallel \ell_3$

## SOLUTIONS TO CLASS EXERCISES

1. $.7^\circ = 7 \times 6$ minutes

$= 42$ minutes

$32.7^\circ = 32^\circ 42'$

2. 60 minutes = 1 degree

$$6.0\overline{\smash{)}2.70}^{\,.45}$$

$56^\circ 27' = 56.45^\circ$

3. acute

4. 
$$\begin{array}{r} 89^\circ 60' \\ 90^\circ \\ -85^\circ 30' \\ \hline 4^\circ 30' \end{array}$$

5. $94^\circ 30'$

6. The measure of the supplement of $\angle A$ is four times that of the complement of $\angle A$.

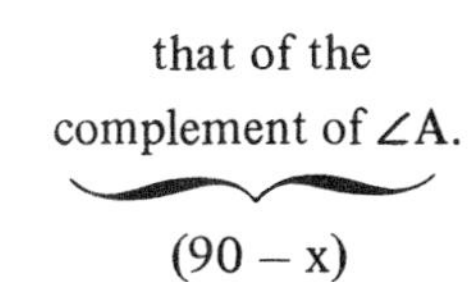

$$180 - x = 4\,(90 - x)$$

$$180 - x = 360 - 4x$$

$$3x = 180$$

$$x = 60$$

$\angle A = 60^\circ$

7. $105^\circ$    8. $75^\circ$    9. $105^\circ$    10. $75^\circ$

11. $105^\circ$    12. $75^\circ$    13. $105^\circ$

14. $\dfrac{4 \text{ in.}}{x \text{ in.}} = \dfrac{12 \text{ in.}}{15 \text{ in.}}$

$60 = 12x$

$5 = x$

BC = 5 in.

NAME

CLASS

ANSWERS

## HOME EXERCISES

In Exercises 1-4, use a protractor to draw each angle.

1. $30°$
2. $75°$
3. $90°$
4. $150°$

In Exercises 5-10, convert into degrees and minutes.

5. $.2°$
6. $40.2°$
7. $19.5°$
8. $72.9°$
9. $100.8°$
10. $120.05°$

In Exercises 11-16, write in decimal notation.

11. $5°30'$
12. $12°12'$
13. $39°48'$
14. $127°36'$
15. $80°3'$
16. $118°57'$

5. ________
6. ________
7. ________
8. ________
9. ________
10. ________
11. ________
12. ________
13. ________
14. ________
15. ________
16. ________

ANSWERS

17. ____________

18. ____________

19. ____________

20. ____________

21. ____________

22. ____________

23. ____________

24. ____________

25. ____________

26. ____________

27. ____________

28. ____________

29. ____________

30. ____________

31. ____________

32. ____________

33. ____________

34. ____________

35. ____________

36. ____________

37. ____________

In Exercises 17-20, classify as either a right angle, a straight angle, an acute angle, or an obtuse angle.

17. $79^\circ$ 18. $180^\circ$ 19. $91^\circ$ 20. $90^\circ 10'$

In Exercises 21-26, find the measure of the complement of each angle.

21. $10^\circ$ 22. $80^\circ$ 23. $45^\circ$

24. $7^\circ 37'$ 25. $19.2^\circ$ 26. $19^\circ 2'$

In Exercises 27-32, find the measure of the supplement of each angle.

27. $50^\circ$ 28. $130^\circ$ 29. $37.4^\circ$

30. $112^\circ 15'$ 31. 39 minutes 32. a right angle

33. An angle measures twice as much as its complement. Find the measure of the angle.

34. An angle measures one-fifth as much as its complement. Find the measure of the angle.

35. An angle measures four-fifths as much as its supplement. Find the measure of the angle.

36. An angle measures $30^\circ$ more than its supplement. Find the measure of the angle.

37. The supplement of angle A measures three times as much as the complement of angle A. Find the measure of angle A.

NAME

CLASS

ANSWERS

38. The complement of angle B measures one-tenth as much as the supplement of angle B. Find the measure of angle B.

38. ____________

39. In Figure 3.17, $\angle BAD = 45^\circ$ and AC is the bisector of $\angle BAD$. Find the measure of $\angle 1$.

39. ____________

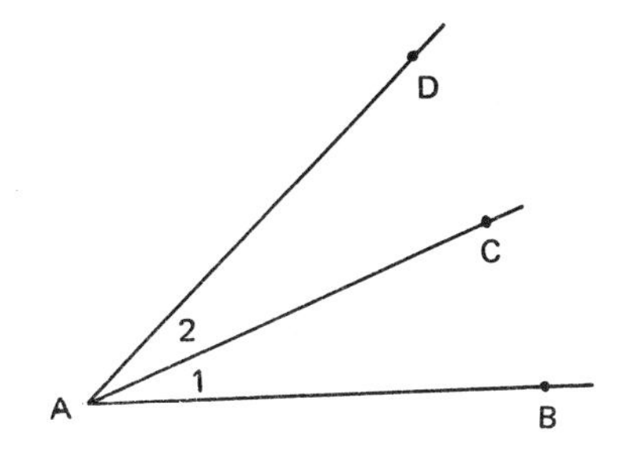


FIGURE 3.17

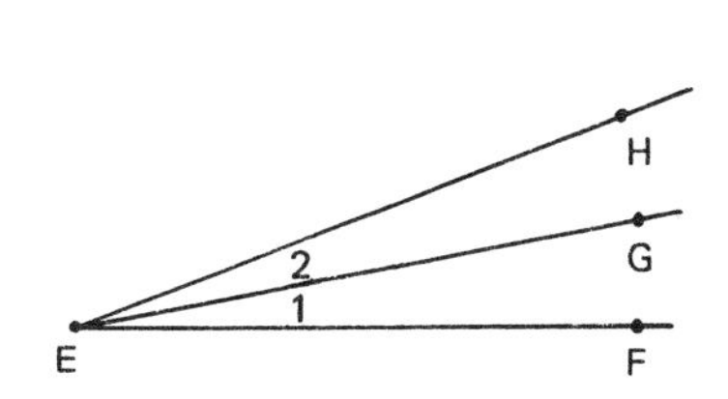


FIGURE 3.18

40. ____________

41. ____________

42. ____________

40. In Figure 3.18, $\angle FEH = 20^\circ 28'$ and EG is the bisector of $\angle FEH$. Find the measure of $\angle 2$.

43. ____________

Exercises 41-43 refer to Figure 3.19. Find the measure of each angle.

41. $\angle 2$ 42. $\angle 3$ 43. $\angle 4$

44. ____________

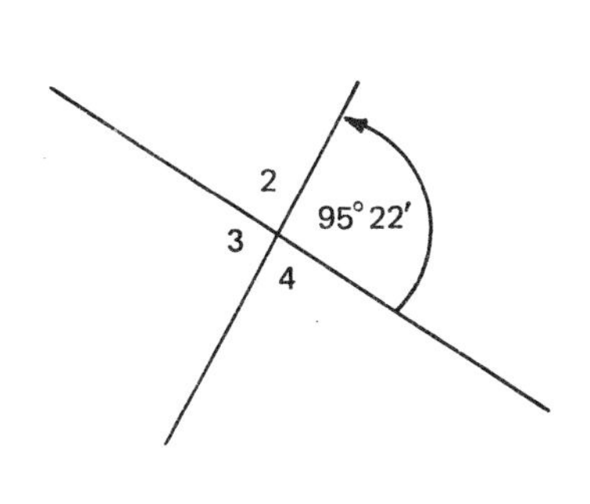


FIGURE 3.19

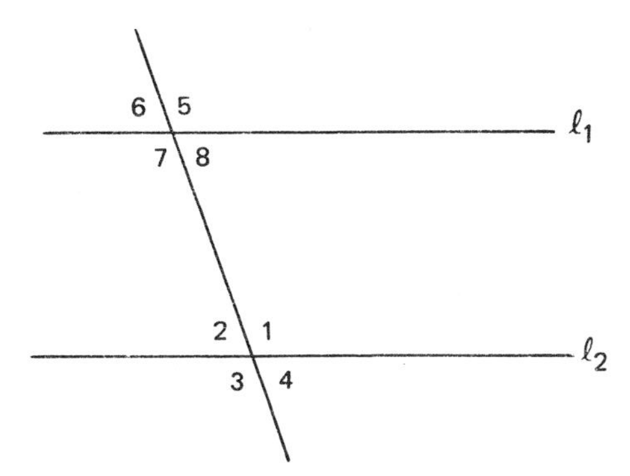


FIGURE 3.20

45. ____________

46. ____________

47. ____________

48. ____________

Exercises 44-50 refer to Figure 3.20. Assume $\ell_1 \parallel \ell_2$ and $\angle 5 = 110^\circ$. Find the measure of each angle.

49. ____________

44. $\angle 1$ 45. $\angle 2$ 46. $\angle 3$ 47. $\angle 4$

48. $\angle 6$ 49. $\angle 7$ 50. $\angle 8$

50. ____________

ANSWERS

51. ______
52. ______
53. ______
54. ______
55. ______
56. ______
57. ______
58. ______
59. ______
60. ______

Exercises 51-60 refer to Figure 3.21.
Assume $\ell_1 \parallel \ell_2 \parallel \ell_3 \parallel \ell_4$.
Also, $\angle BAE = 88°$, $\angle EFB = 117°$, AB = 3 inches, BC = 6 inches, EF = 4 inches, FH = 24 inches.

Find the measure of each angle.

51. $\angle CBF$
52. $\angle GCB$
53. $\angle HDC$
54. $\angle AEF$
55. $\angle CGH$
56. $\angle GHD$

Find the length of each line segment.

57. FG
58. BD
59. CD
60. GH

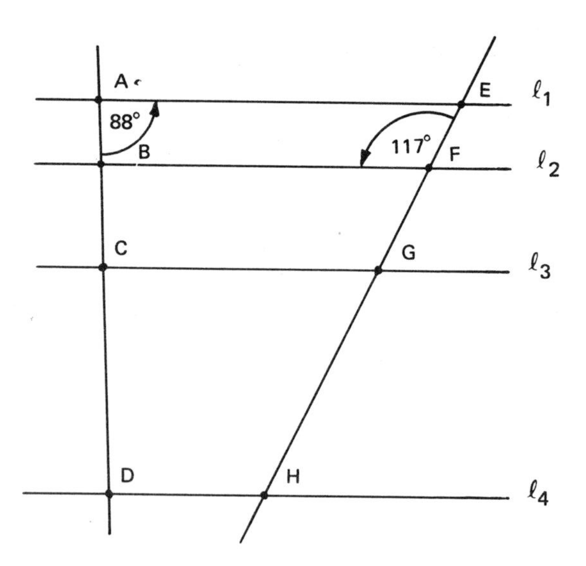


FIGURE 3.21

## 3.2 More on Triangles

### ANGLES AND TRIANGLES

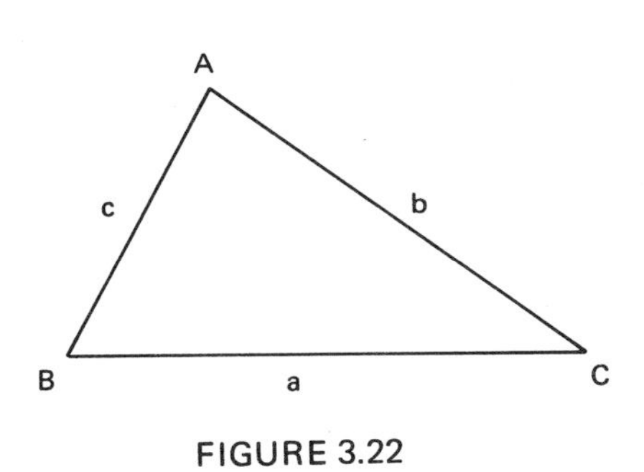


FIGURE 3.22

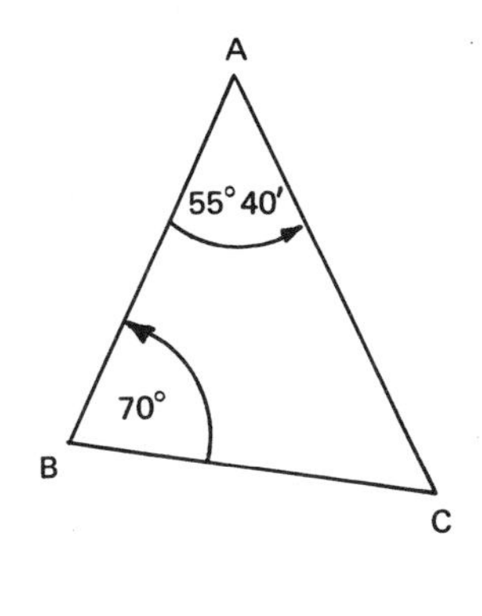


FIGURE 3.23

$\angle C = 180° - (55°40' + 70°)$
$= 180° - 125°40'$
$= 54°20°$

In Figure 3.22, the three vertices of the triangle are labeled A, B and C, and the triangle is denoted by

$$\triangle ABC.$$

The side opposite $\angle A$ is labeled a, the side opposite $\angle B$ is labeled b, and the side opposite $\angle C$ is labeled c.

*The sum of the measures of the angles of a triangle is 180°*. Thus, in $\triangle ABC$ of Figure 3.23, $\angle A = 55°40'$, $\angle B = 70°$. It follows that $\angle C$ is the supplement of $\angle A + \angle B$, or $54°20'$.

Recall that a *right triangle* is a triangle in which one angle is a right angle. An *acute triangle* is a triangle in which all three angles are acute. An *obtuse triangle* is a triangle in which one angle is obtuse.

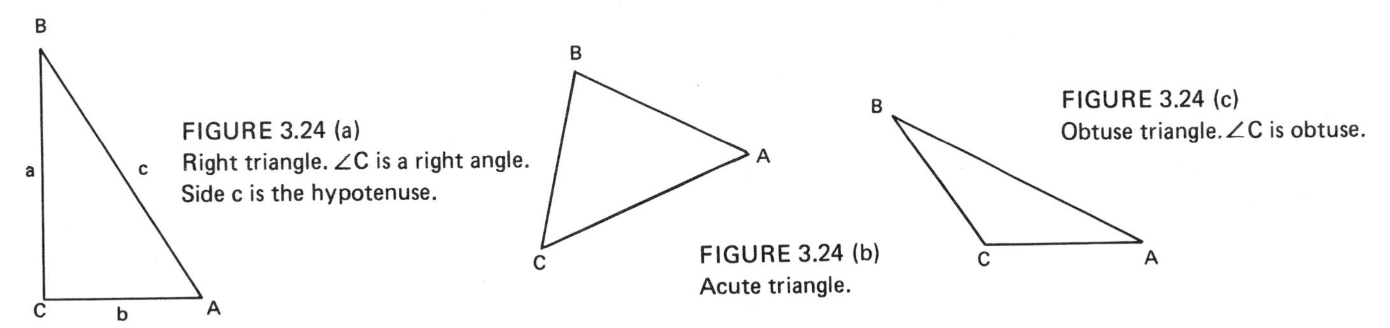


FIGURE 3.24 (a)
Right triangle. $\angle C$ is a right angle.
Side c is the hypotenuse.

FIGURE 3.24 (b)
Acute triangle.

FIGURE 3.24 (c)
Obtuse triangle. $\angle C$ is obtuse.

### CLASS EXERCISES

1. (a) In Figure 3.25, determine the measure of $\angle C$.

   (b) Is this a right triangle, an acute triangle, or an obtuse triangle?

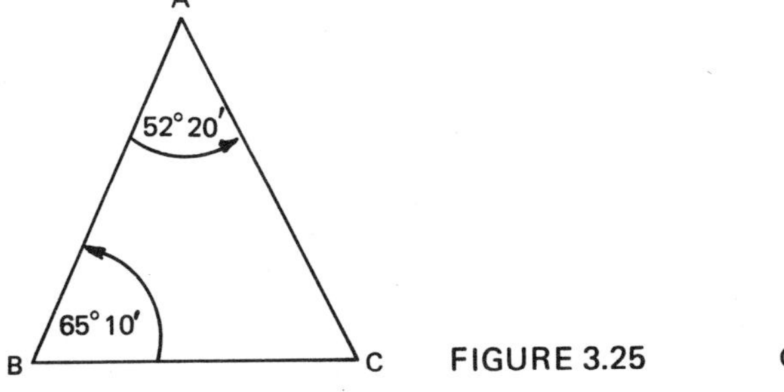


FIGURE 3.25

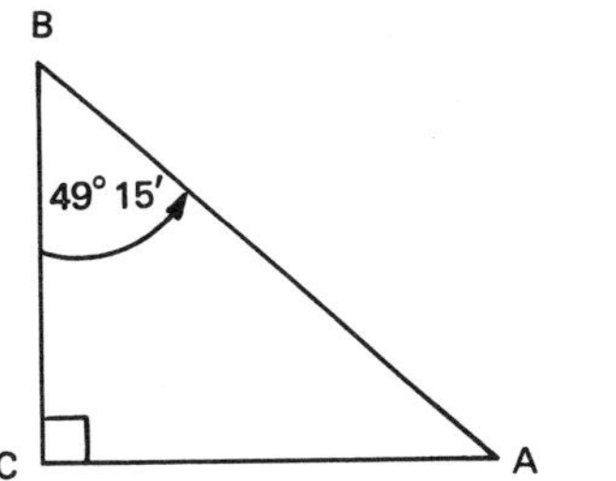


FIGURE 3.26

2. In Figure 3.26, $\angle C$ is a right angle. Determine the measure of $\angle A$.

## ISOSCELES TRIANGLES

An *isosceles triangle* is a triangle in which exactly two sides are equal in length. The equal sides of an isosceles triangle are called its *legs* and the third side is usually called the *base.* The angles opposite the legs are also equal (in measure) and are called the *base angles.* Figure 3.27 shows an isosceles triangle with legs b and c. Note that $\angle B = \angle C$. Also, the height corresponding to base a *bisects* a, that is, divides a into 2 equal line segments. Thus, BD = DC

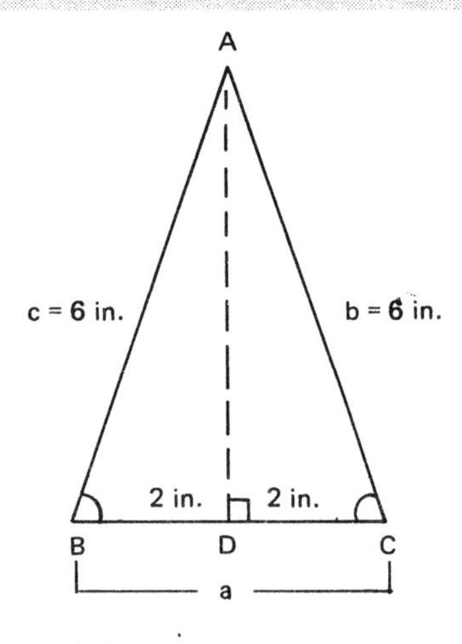


FIGURE 3.27 b = c. Therefore, $\angle B = \angle C$. Also, BD = DC

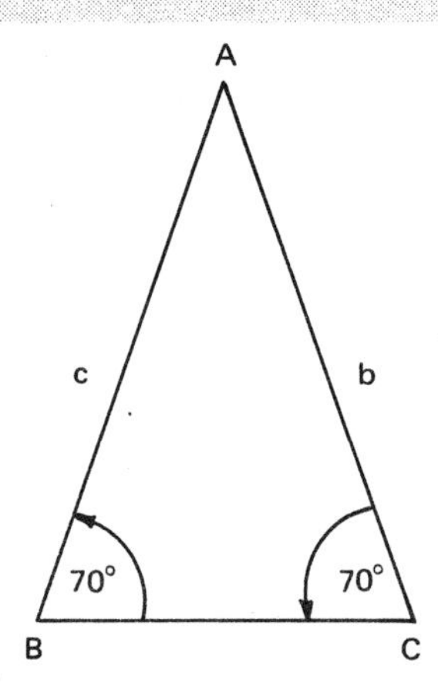


FIGURE 3.28 $\angle B = \angle C$. Therefore, b = c

If exactly two angles of a triangle are equal, the triangle is isosceles. Thus, Figure 3.28 depicts an isosceles triangle with legs b and c.

In an *isosceles right triangle,* the equal base angles must each measure 45°. In fact, if x is the number of degrees in each of these equal angles, then

$$x + x + 90 = 180$$

$$2x = 90$$

$$x = 45$$

Thus, an *isosceles* right triangle is often called a 45°, 45°, 90°- triangle.

EXAMPLE 1. Find (a) the hypotenuse and (b) the area of $\triangle ABC$.

SOLUTION. (a) $\angle B = 45°$, and thus, $\triangle ABC$ is an isosceles right triangle. It follows that b = 5 inches. By the Pythagorean Theorem,

$$a^2 + b^2 = c^2$$

$$(5 \text{ in.})^2 + (5 \text{ in.})^2 = c^2$$

$$25 \text{ in.}^2 + 25 \text{ in.}^2 = c^2$$

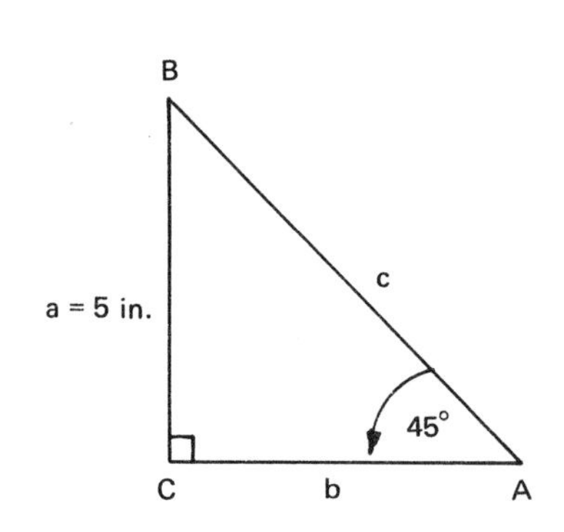


FIGURE 3.29

$$50 \text{ in.}^2 = c^2$$

$$\sqrt{50} \text{ in.} = c$$

$$5\sqrt{2} \text{ in.} = c$$

(b) The area of a triangle is given by the formula

$$\text{Area} = \frac{\text{base} \cdot \text{height}}{2}$$

$$A = \frac{5 \text{ in.} \times 5 \text{ in.}}{2}$$

$$A = \frac{25}{2} \text{ in.}^2$$

Example 1 (a) illustrates that *in an isosceles right triangle with legs a and b and hypotenuse c, the sides are in the ratio of*

$$\begin{matrix} 1 & : & 1 & : & \sqrt{2} \\ \uparrow & & \uparrow & & \uparrow \\ a & & b & & c \end{matrix}$$

In other words, sides a and b are of equal length and

$$\text{length of } c = \sqrt{2} \times (\text{length of } a)$$

## CLASS EXERCISES

3. Consider Figure 3.30. (a) Show that $\Delta ABC$ is an isosceles triangle. (b) Find the length of side BC.

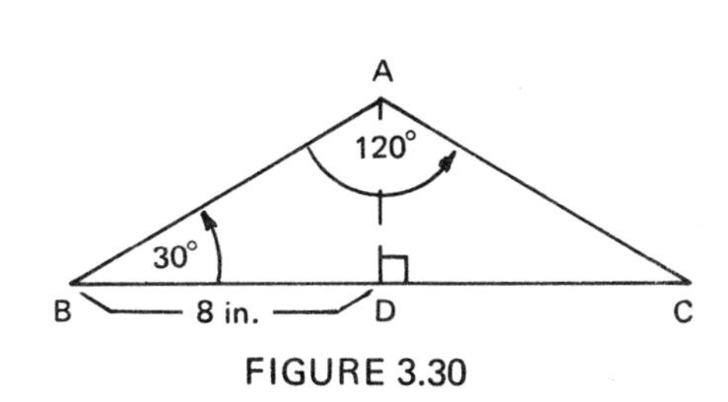


FIGURE 3.30

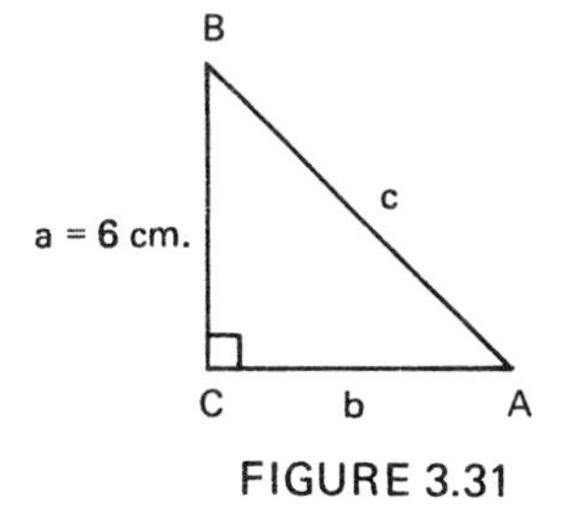


FIGURE 3.31

4. In Figure 3.31, suppose $\Delta ABC$ is an isosceles right triangle. Find (a) its hypotenuse and (b) its area.

### EQUILATERAL TRIANGLES

An *equilateral triangle* is one in which all three side lengths (and all three angle measurements) are equal. Each angle of an equilateral triangle measures 60°. Each height bisects the corresponding base as well as the corresponding angle. [See Figure 3.32(a).]

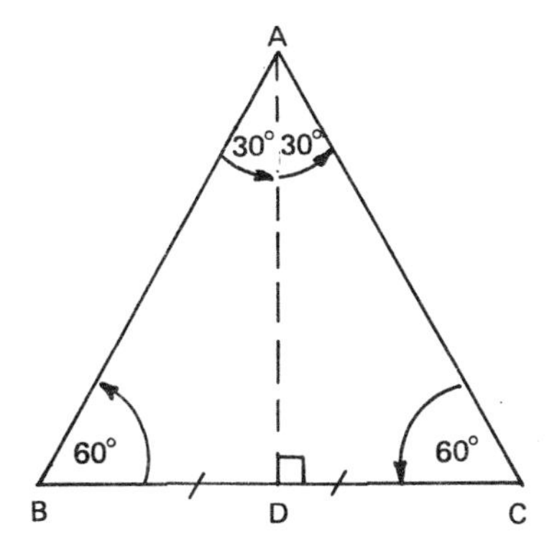


FIGURE 3.32 (a)
An equilateral triangle.
$\angle DAB = \angle CAD = 30°$

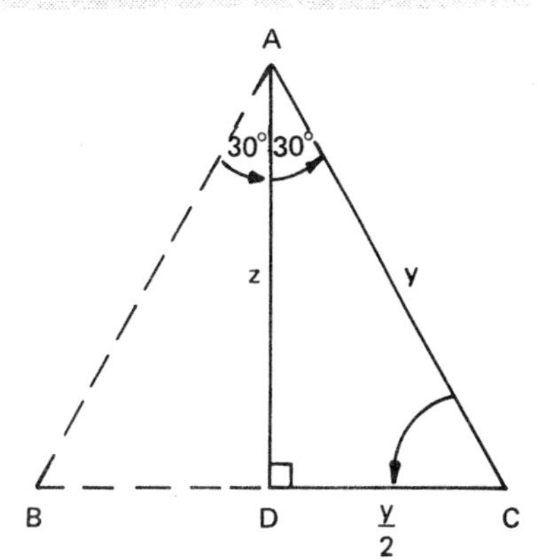


FIGURE 3.32 (b)

In Figure 3.32(b), $\Delta ADC$ is known as a 30°, 60°, 90°-triangle. Note that DC, the side opposite the 30°-angle, is ½ as long as side AC, because

$$DC = \frac{1}{2} BC = \frac{1}{2} AC$$

Let $z = AD$

and $y = AC$.

Then $\frac{y}{2} = DC$,

and by the Pythagorean Theorem,

$$z^2 + \left(\frac{y}{2}\right)^2 = y^2$$

$$z^2 + \frac{y^2}{4} = y^2$$

$$z^2 = \frac{3y^2}{4}$$

$$z = \frac{\sqrt{3}y}{2}$$

Thus, the ratio of the side lengths of a 30°, 60°, 90°-triangle is indicated in Figure 3.33. Here, let x be the side opposite the 30°-angle. Then

$$x : y : z$$

as

$$1 : 2 : \sqrt{3}$$

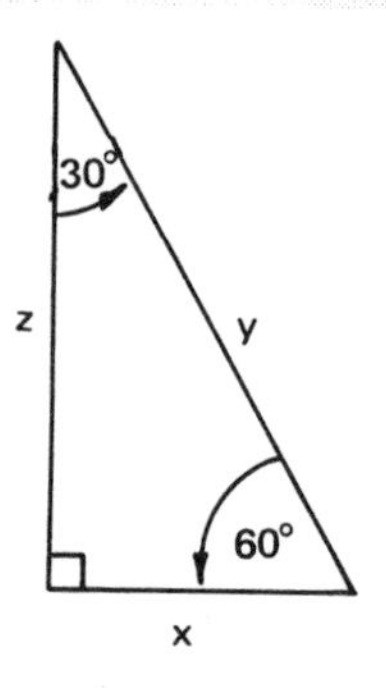


FIGURE 3.33 (a)

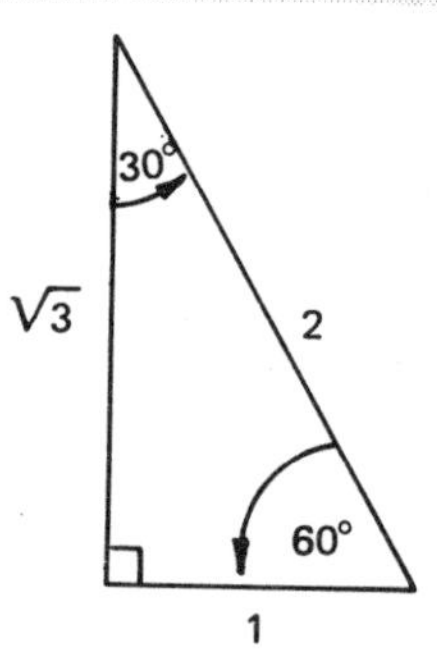


FIGURE 3.33 (b)

EXAMPLE 2. Consider the triangle in Figure 3.34. Find the length of sides b and c.

SOLUTION. This is a 30°, 60°, 90°-triangle.

(a)

$$\frac{a}{b} = \frac{\sqrt{3}}{1}$$

$$\frac{10 \text{ in.}}{b} = \sqrt{3}$$

$$10 \text{ in.} = b \cdot \sqrt{3}$$

$$\frac{10}{\sqrt{3}} \text{in.} = b$$

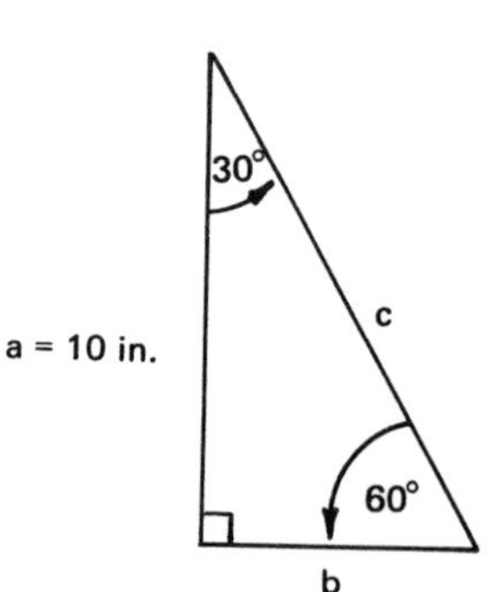


FIGURE 3.34

It is easier to work with a fraction whose denominator is a rational number. *Rationalize the denominator* on the left by multiplying the fraction by

$$\frac{\sqrt{3}}{\sqrt{3}},$$

which equals 1, so that the new fraction equals the old one.

$$\frac{10}{\sqrt{3}} \cdot \frac{3}{\sqrt{3}} \text{ in.} = b$$

$$\frac{10\sqrt{3}}{3} \text{ in.} = b$$

(b)

$$c = 2b$$

$$= \frac{2 \cdot 10\sqrt{3}}{3} \text{ in.}$$

$$= \frac{20\sqrt{3}}{3} \text{ in.}$$

Now let us return to an equilateral triangle with side length s, as in Figure 3.35. Observe that $\triangle ADC$ is a $30°, 60°, 90°$-triangle. Because

$$DC = \frac{s}{2},$$

it follows that

$$h = \frac{s\sqrt{3}}{2}.$$

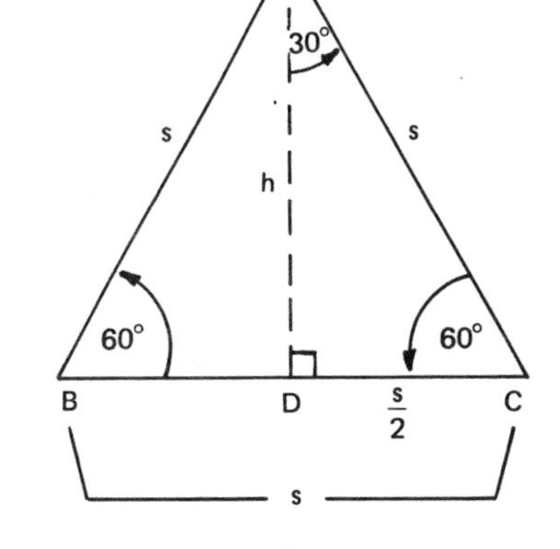


FIGURE 3.35

To find the area, A, of this equilateral triangle, recall that

$$A = \frac{1}{2} \text{ base} \cdot \text{height}$$

$$A = \frac{1}{2} \cdot s \cdot \frac{s\sqrt{3}}{2}$$

$$= \frac{s^2\sqrt{3}}{4}$$

Thus, *the area, A, of an equilateral triangle with side length s is given by the formula*

$$A = \frac{s^2\sqrt{3}}{4}$$

EXAMPLE 3. Find (a) the height and (b) the area of an equilateral triangle with side length 6 decimeters.

SOLUTION. (a)

$$h = \frac{s\sqrt{3}}{2} \text{ dm.}$$

$$= \frac{6\sqrt{3}}{2} \text{ dm.}$$

$$= 3\sqrt{3} \text{ dm.}$$

The height is $3\sqrt{3}$ decimeters.

(b)

$$A = \frac{s^2\sqrt{3}}{4}$$

$$= \frac{6^2\sqrt{3}}{4} \text{ dm.}^2$$

$$= \frac{36\sqrt{3}}{4} \text{ dm.}^2$$

$$= 9\sqrt{3} \text{ dm.}^2$$

The area is $9\sqrt{3}$ square decimeters.

EXAMPLE 4. Find (a) the side length and (b) the height of an equilateral triangle whose area is $25\sqrt{3}$ square feet.

SOLUTION. (a)

$$A = \frac{s^2\sqrt{3}}{4}$$

$$25\sqrt{3} \text{ ft.}^2 = \frac{s^2\sqrt{3}}{4}$$

$$100 \text{ ft.}^2 = s^2$$

$$10 \text{ ft.} = s$$

The side length is 10 feet.

(b)

$$h = \frac{s\sqrt{3}}{2}$$

$$h = \frac{10\sqrt{3}}{2} \text{ ft.}$$

$$h = 5\sqrt{3} \text{ ft.}$$

The height is $5\sqrt{3}$ feet.

## CLASS EXERCISES

If necessary, leave your answers in terms of $\sqrt{3}$.

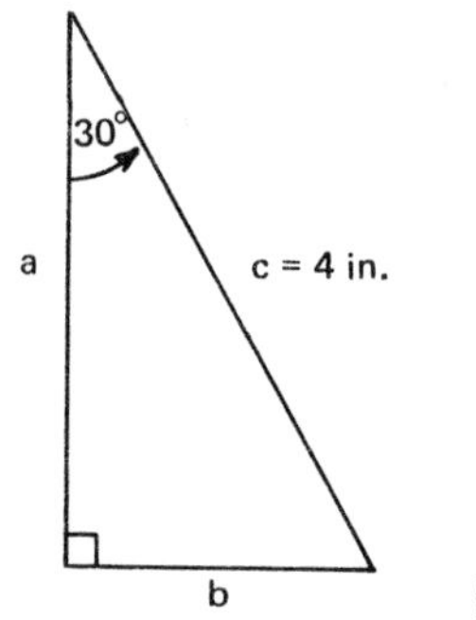


FIGURE 3.36 (a)

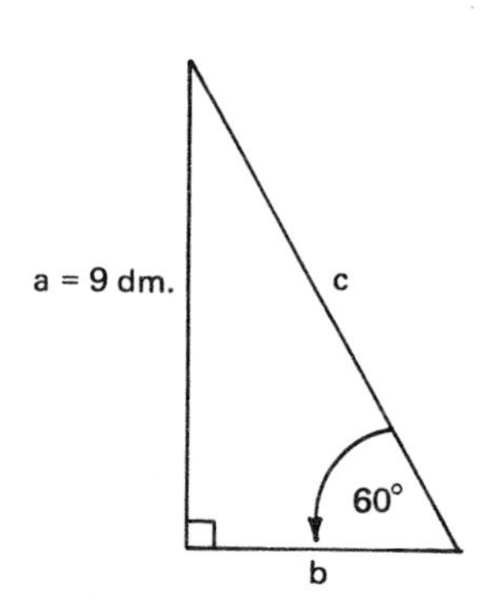


FIGURE 3.36 (b)

5. Consider the triangle in Figure 3.36 (a). Find the lengths of sides a and b.

6. Consider the triangle in Figure 3.36 (b). Find the lengths of sides b and c.

7. Find (a) the height and (b) the area of an equilateral triangle with side length 5 meters.

8. Find (a) the side length and (b) the height of an equilateral triangle whose area is $100\sqrt{3}$ square centimeters.

## SIMILAR TRIANGLES

Similar figures have the same shape. Recall that *two triangles are similar if the lengths of their corresponding sides are proportional.* Similarity of triangles can also be described in terms of angles. *Two triangles are similar if their corresponding angles are equal (in measure). Whenever corresponding side lengths are proportional, corresponding angles are equal, and vice versa.*

*If two angles of one triangle are equal to two angles of another triangle, the third angles must also be equal and the triangles are then similar.* (See Class Exercise 9.)

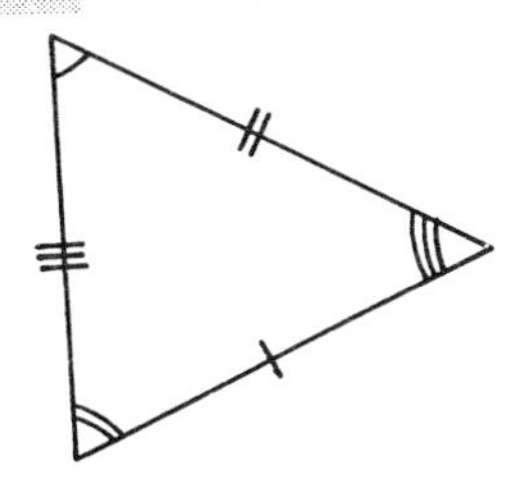

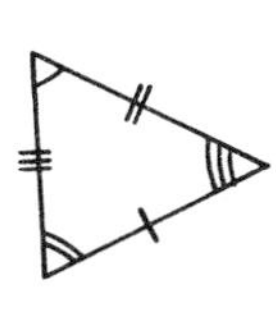

FIGURE 3.37 Similar triangles.
Corresponding sides and corresponding angles are indicated.

EXAMPLE 5. In Figure 3.38 (a), assume AB || DE.

(a) Show that ΔABC is similar to ΔCDE.

(b) What is the ratio of similarity?

(c) Find the length of AB.

(d) Find the length of CE.

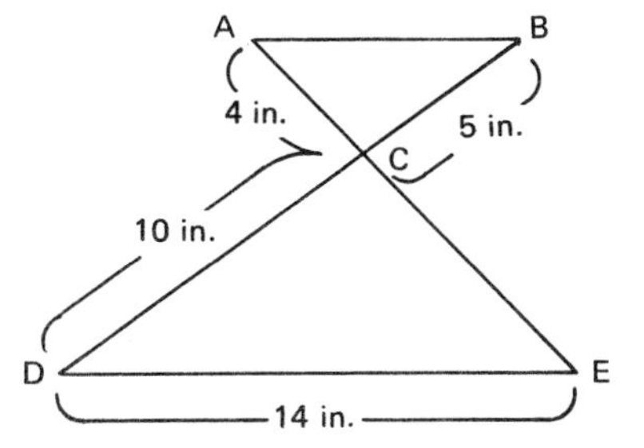


FIGURE 3.38 (a)

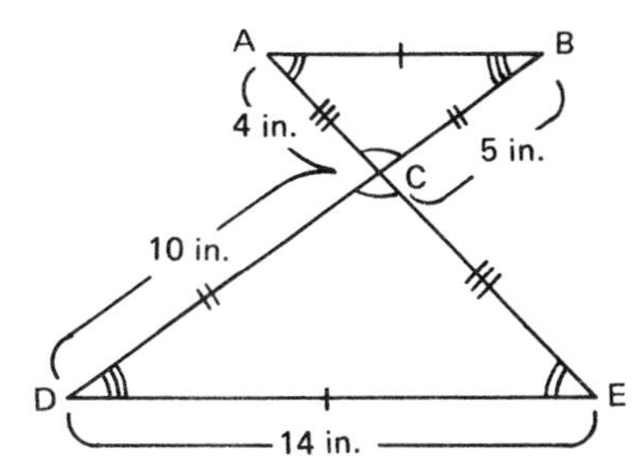


FIGURE 3.38 (b)

SOLUTION. (a) $\angle BCA = \angle DCE$ because vertical angles are equal. $\angle A = \angle E$ and $\angle B = \angle D$ because AB || DE and $\angle A$ and $\angle E$ are alternate interior angles, as are $\angle B$ and $\angle D$. The two triangles are therefore similar.

In Figure 3.38 (b), corresponding angles are indicated. Note that the sides opposite corresponding angles also correspond.

(b)

$$\text{Ratio of Similarity} = \frac{DC}{BC} = \frac{10 \text{ in.}}{5 \text{ in.}} = 2$$

(c)

$$\frac{DE}{AB} = 2$$

$$\frac{14 \text{ in.}}{AB} = 2$$

$$\frac{14 \text{ in.}}{2} = AB$$

$$7 \text{ in.} = AB$$

(d)

$$\frac{CE}{AC} = 2$$

$$\frac{CE}{4 \text{ in.}} = 2$$

$$CE = 8 \text{ in.}$$

## CLASS EXERCISES

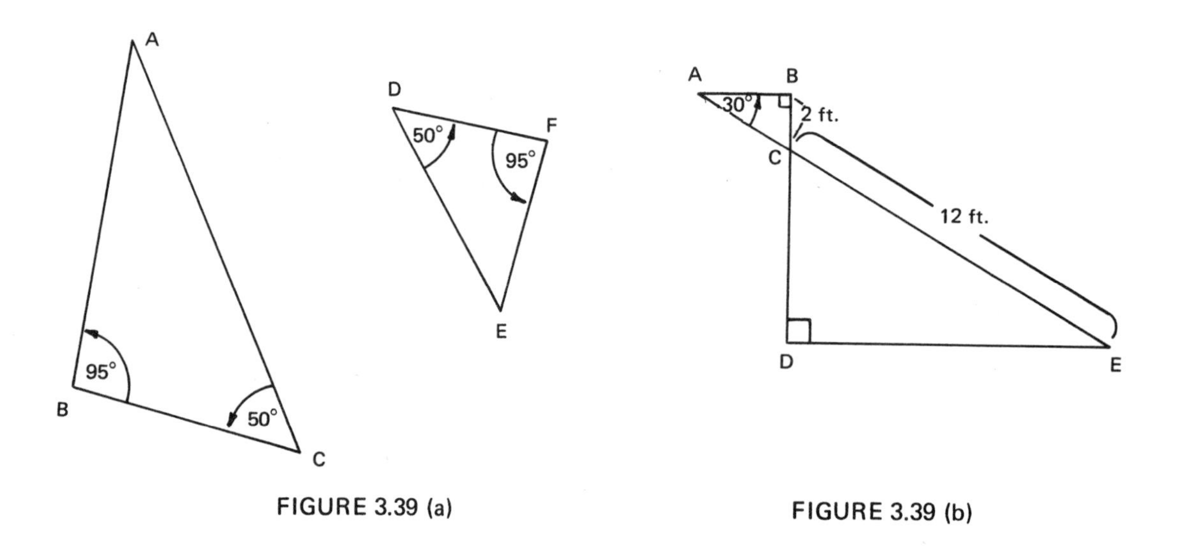


FIGURE 3.39 (a)

FIGURE 3.39 (b)

9. In Figure 3.39 (a) show that $\Delta ABC$ is similar to $\Delta DEF$.

10. In Figure 3.39 (b):
 (a) Show that $\Delta$ABC is similar to $\Delta$CDE. (b) Find the ratio of similarity. (c) Find area $\Delta$ABC.

## PYTHAGOREAN THEOREM

Recall that the *Pythagorean Theorem* states that in a right triangle with base b, height h, and hypotenuse c,

$$b^2 + h^2 = c^2.$$

Here is a second proof of the Pythagorean Theorem, using the preceding material on similarity of triangles. In Figure 3.40, $\Delta$ABC and $\Delta$BCD are right triangles that share a common angle, $\angle$B. The remaining angles must have equal measure. Thus,

$$\angle A = \angle BCD,$$

and $\Delta$ABC and $\Delta$BCD are similar. Corresponding sides are proportional. Therefore,

$$\frac{b}{c} = \frac{x}{b}$$

and by cross-multiplying,

(1) $b^2 = cx$

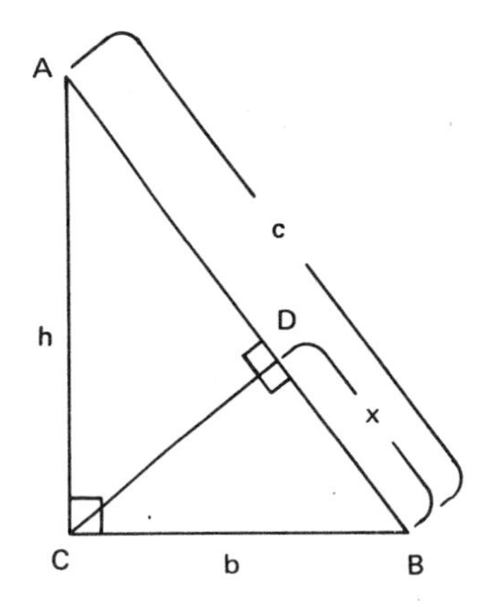


FIGURE 3.40

Also, $\Delta$ABC and $\Delta$ACD are right triangles that share a common angle, $\angle$A. Thus,

$$\angle B = \angle DCA,$$

and $\Delta$ABC and $\Delta$ACD are similar. Corresponding sides are proportional. Therefore,

$$\frac{h}{c} = \frac{c - x}{h},$$

and by cross-multiplying,

(2) $$h^2 = c^2 - cx$$

Add the left side and the right sides of equations (1) and (2) to obtain

$$b^2 + h^2 = cx + (c^2 - cx)$$

Thus, $$b^2 + h^2 = c^2$$

This, of course, is the conclusion of the Pythagorean Theorem.

## SOLUTIONS TO CLASS EXERCISES

1. (a) $\angle A + \angle B = 52°20' + 65°10'$

$= 117°30'$

$\angle C = 180° - 117°30'$

$\angle C = 62°30'$

(b) an acute triangle

2. $\angle A = 90° - 49°15'$

$\angle A = 40°45'$

3. (a) $\angle A + \angle B = 120° + 30°$

$= 150°$

$\angle C = 180° - 50°$

$= 30°$

Thus, $\angle C = \angle B$, and $\triangle ABC$ is isosceles.

(b) AD bisects BC. Thus, BC = 2 · BD. Therefore, BC = 16 inches.

4. (a) $c = a\sqrt{2}$

$= 6\sqrt{2}$ cm.

(b) $A = \frac{6 \text{ cm.} \times 6 \text{ cm.}}{2}$

$= 18 \text{ cm.}^2$

5. $c = 2b$

4 in. = 2b

2 in. = b

$a = b\sqrt{3}$

$a = 2\sqrt{3}$ in.

6. $\frac{a}{b} = \frac{\sqrt{3}}{1}$

$\frac{9 \text{ dm.}}{b} = \sqrt{3}$

$9 \text{ dm.} = b\sqrt{3}$

$\frac{9}{\sqrt{3}} \text{dm.} = b$

$\frac{9\sqrt{3}}{\sqrt{3}\sqrt{3}} \text{dm.} = b$

$\frac{9\sqrt{3}}{3} \text{dm.} = b$

$3\sqrt{3} \text{ dm.} = b$

$c = 2b$

$c = 2 \cdot 3\sqrt{3}$ dm.

$c = 6\sqrt{3}$ dm.

7. (a) $h = \frac{s\sqrt{3}}{2}$

$h = \frac{5\sqrt{3}}{2}$ m.

(b) $A = \frac{s^2\sqrt{3}}{4}$

$A = \frac{(5 \text{ m.})^2\sqrt{3}}{4}$

$A = \frac{25\sqrt{3}}{4} \text{m.}^2$

8. (a) $A = \frac{s^2 \sqrt{3}}{4}$

$100 \sqrt{3}\text{ cm.}^2 = \frac{s^2 \sqrt{3}}{4}$

$400 \text{ cm.}^2 = s^2$

$20 \text{ cm.} = s$

(b) $h = \frac{s \sqrt{3}}{2}$

$h = \frac{20 \text{ cm.} \times \sqrt{3}}{2}$

$h = 10 \sqrt{3} \text{ cm.}$

9. $\angle B = \angle F = 95°$

$\angle C = \angle D = 50°$

$\angle A = 180° - (95° + 50°)$

$= 180° - 145°$

$= 35°$

And $\angle E = 180° - (95° + 50°)$

$= 35°$

Thus, $\angle A = \angle E$

Therefore, $\triangle ABC$ is similar to $\triangle DEF$.

10. (a) $\angle B = \angle D = 90°$

Therefore, AB || DE because alternate interior angles are equal.

$\angle A = \angle E = 30°$ (alternate interior angles)

$\angle BCA = \angle DCE = 60°$

Therefore, $\triangle ABC$ is similar to $\triangle CDE$.

(b) AC = 4 ft. (30°, 60°, 90°-triangle)

Ratio of similarity $= \frac{12 \text{ ft.}}{4 \text{ ft.}} = 3$

(c) $AB = 2\sqrt{3}$ ft.

Area $\triangle ABC = \frac{2 \text{ ft.} \times 2\sqrt{3} \text{ ft.}}{2} = 2\sqrt{3} \text{ ft.}^2$

NAME ____________

CLASS ____________

ANSWERS

## HOME EXERCISES

In Exercises 1-10, find the measure of the third angle of each triangle.

1.

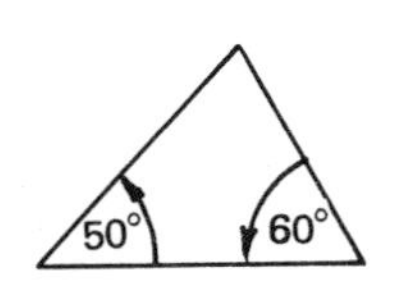


FIGURE 3.41

2.

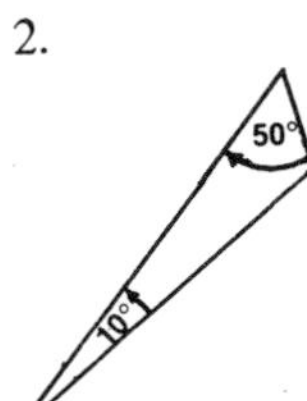


FIGURE 3.42

3.

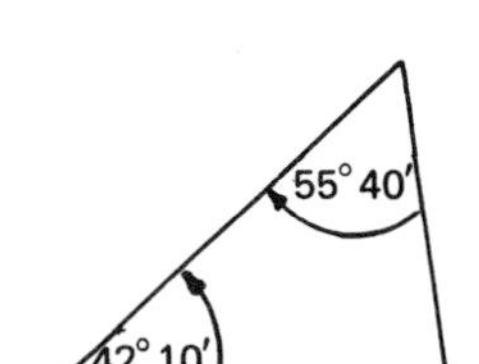


FIGURE 3.43

4.

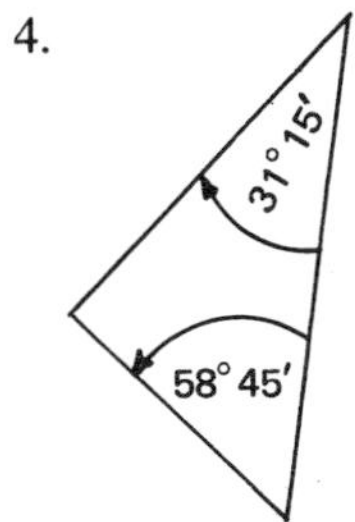


FIGURE 3.44

5.

110.7°

42.9°

FIGURE 3.45

6.

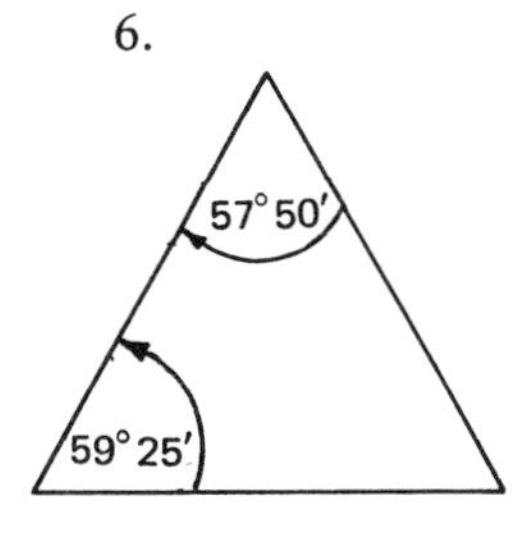


FIGURE 3.46

7.

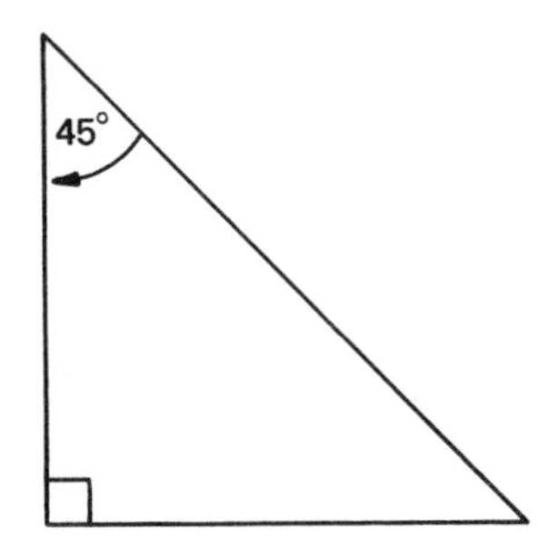


FIGURE 3.47

8.

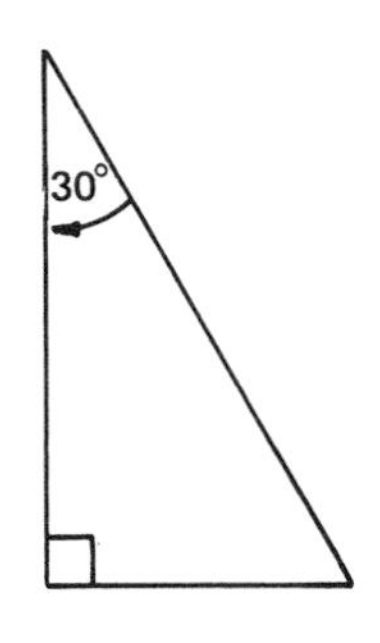


FIGURE 3.48

9.

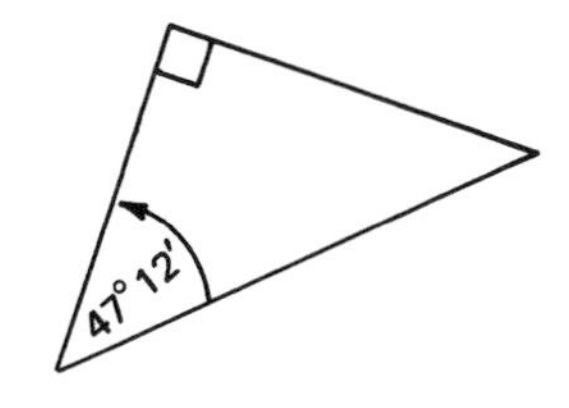


FIGURE 3.49

10.

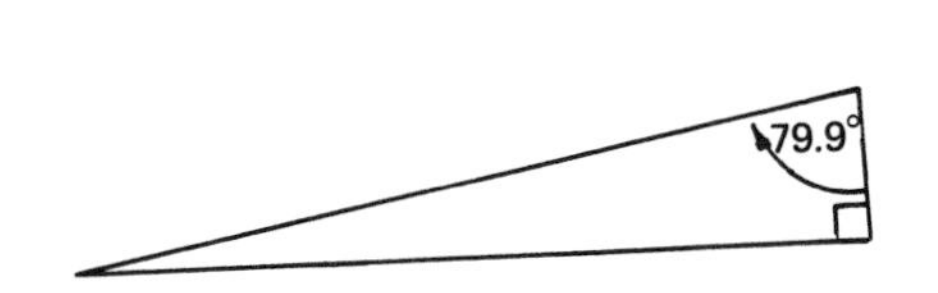


FIGURE 3.50

1. ____________

2. ____________

3. ____________

4. ____________

5. ____________

6. ____________

7. ____________

8. ____________

9. ____________

10. ____________

ANSWERS

11. ____________________

12. ____________________

13. ____________________

14. ____________________

15. (a) 

(b) ____________________

(c) ____________________

(d) ____________________

11. In a right triangle one acute angle measures twice as much as the other. Find the measure of each acute angle.

12. In a right triangle one acute angle measures $10^\circ$ more than the other. Find the measure of each acute angle.

13. The largest angle of a triangle measures $20^\circ$ more than the second largest angle and $25^\circ$ more than the smallest angle. Find the measure of each angle.

14. The largest angle of a triangle measures twice another angle and three times the third angle. Find the measure of each angle.

15. In Figure 3.51, which triangles are isosceles?

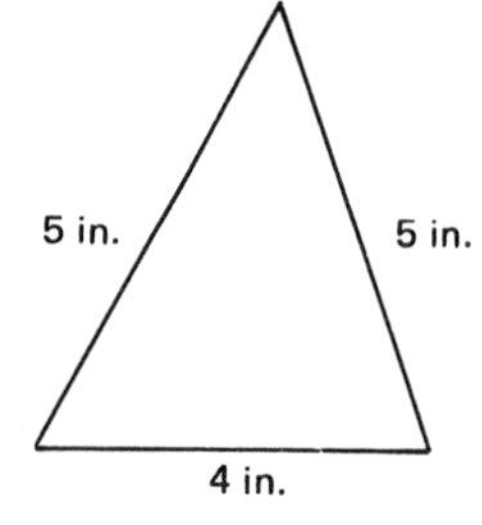


FIGURE 3.51 (a)

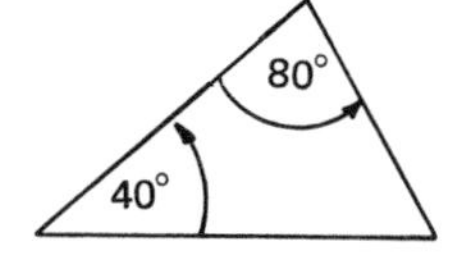


FIGURE 3.51 (b)

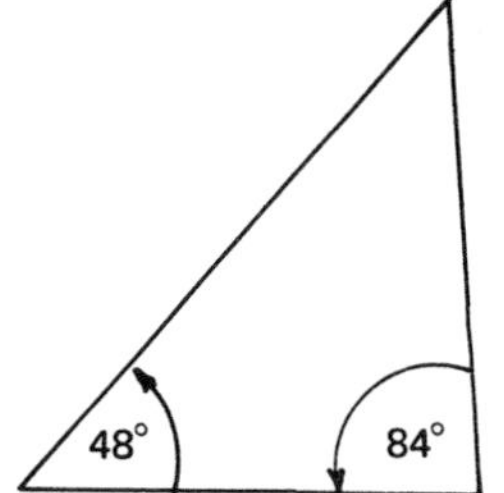


FIGURE 3.51 (c)

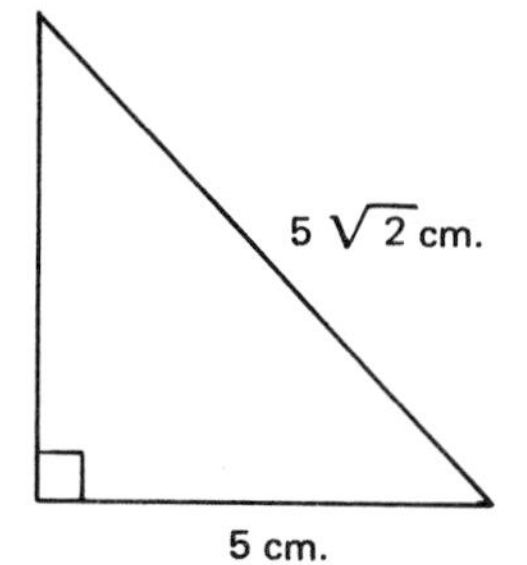


FIGURE 3.51 (d)

NAME

CLASS

ANSWERS

16. Each of the base angles of an isosceles triangle measures $37°30'$. Find the measure of the the third angle.

17. In Figure 3.52, AB = AC and AD is the height corresponding to base BC.

(a) Find the length of BD.

(b) Find the area of $\triangle$ABC.

18. In Figure 3.53, AB = AC and AD is the height corresponding to side BC.

(a) Find the length of base BC.

(b) Find the perimeter of $\triangle$ABC.

(c) Find the perimeter of $\triangle$ABD.

(d) Find the area of $\triangle$ABC.

(e) Find the area of $\triangle$ABD.

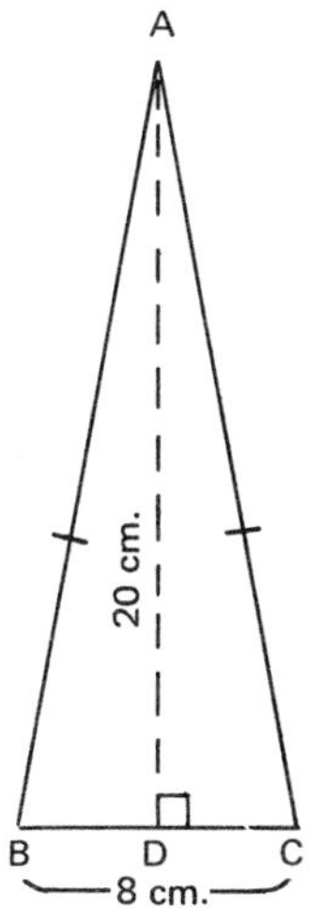

A

20 cm.

B D C

8 cm.

FIGURE 3.52

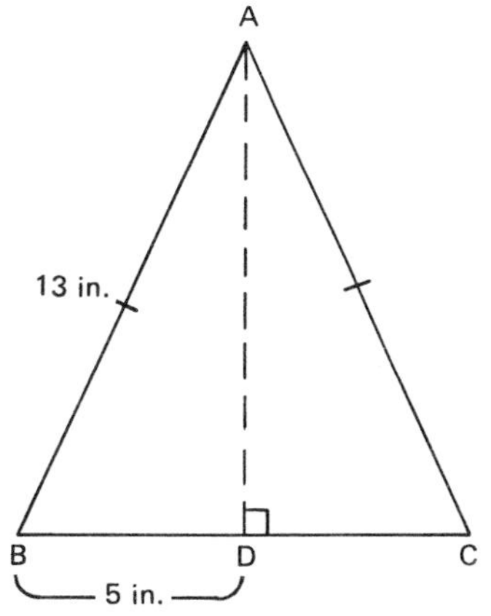

A

13 in.

B D C

5 in.

FIGURE 3.53

16. ____

17. (a) ____

(b) ____

18. (a) ____

(b) ____

(c) ____

(d) ____

(e) ____

ANSWERS

19. In Figure 3.54, suppose $\triangle ABC$ is an isosceles right triangle. Find (a) the hypotenuse and (b) the area.

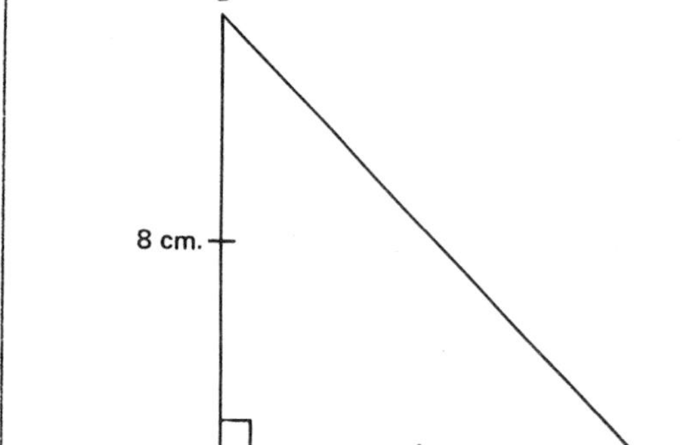


FIGURE 3.54

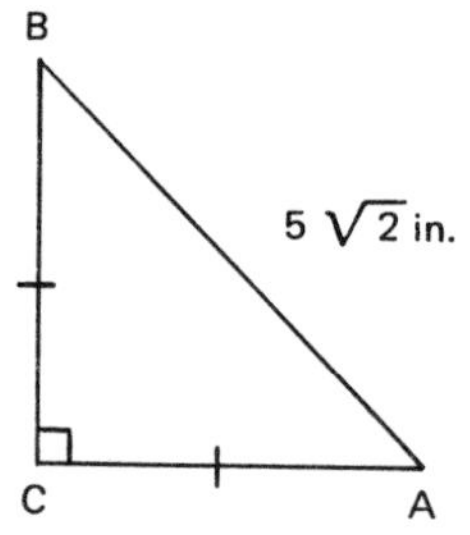

FIGURE 3.55

19. (a)______________

(b)______________

20. In Figure 3.55, suppose $\triangle ABC$ is an isosceles right triangle. Find

(a) the length of each leg and (b) the area.

20. (a)______________

(b)______________

21. Consider the triangle in Figure 3.56. Find the lengths of sides a and b.

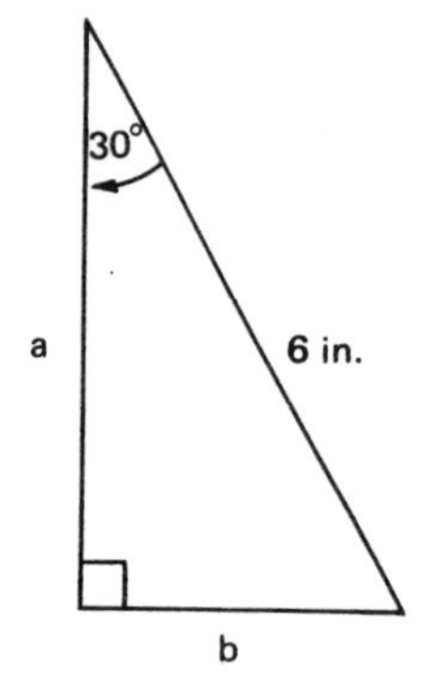


FIGURE 3.56

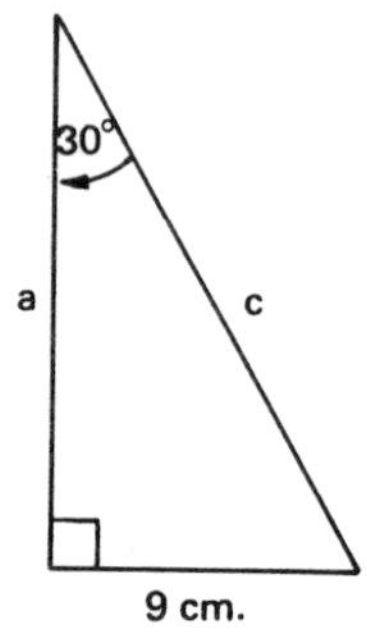

FIGURE 3.57

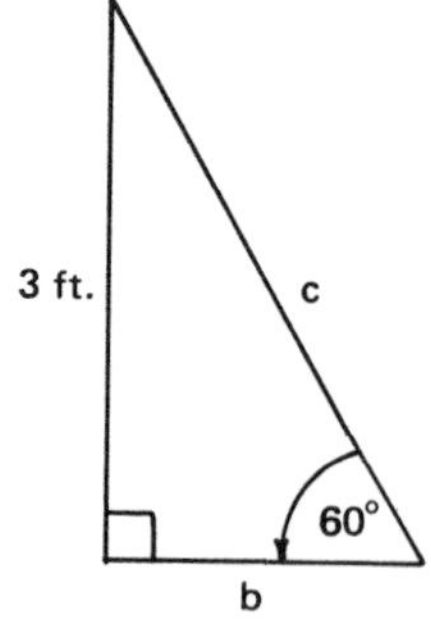


FIGURE 3.58

21. ______________

NAME

CLASS

ANSWERS

22. Consider the triangle in Figure 3.57. Find the lengths of sides a and c.

23. Consider the triangle in Figure 3.58. Find the lengths of sides b and c.

24. Find the area of a $30^\circ, 60^\circ, 90^\circ$-triangle whose hypotenuse is 4 inches.

25. Find (a) the height, (b) the perimeter, and (c) the area of an equalitarial triangle with side length 7 inches.

26. Find (a) the side length and (b) the area of an equilateral triangle with height $10\sqrt{3}$ centimeters.

27. Find (a) the side length and (b) the height of an equilateral triangle with area $9\sqrt{3}$ square inches.

28. An equilateral triangle has perimeter 27 decimeters. Find its area.

Exercises 29-39 refer to Figure 3.59. Find:

29. the measure of $\angle C$

30. the measure of $\angle BAD$

31. the measure of $\angle BAC$

32. the length of DC

33. the length of AC

34. the length of BD

22. ____________

23. ____________

24. ____________

25. (a) ____________

(b) ____________

(c) ____________

26. (a) ____________

(b) ____________

27. (a) ____________

(b) ____________

28. ____________

29. ____________

30. ____________

31. ____________

32. ____________

33. ____________

34. ____________

ANSWERS

35. ______________

36. ______________

37. ______________

38. ______________

39. ______________

40. ______________

41. ______________

42. ______________

43. ______________

44. ______________

45. ______________

46. ______________

47. ______________

35. the length of BC

36. the length of AB

37. the area of $\Delta ADC$

38. the area of $\Delta ABC$

39. the area of $\Delta ABD$

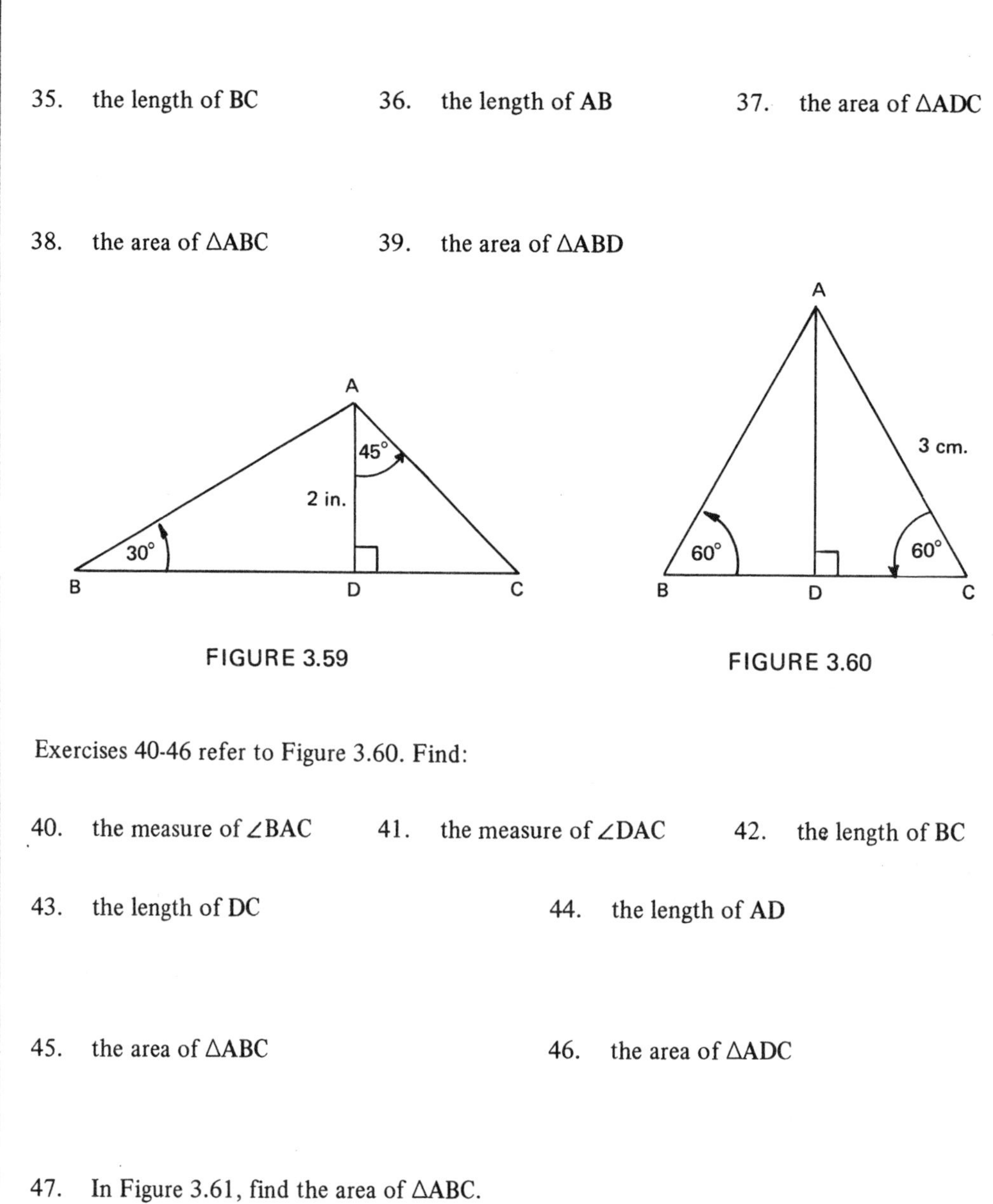


FIGURE 3.59

FIGURE 3.60

Exercises 40-46 refer to Figure 3.60. Find:

40. the measure of $\angle BAC$

41. the measure of $\angle DAC$

42. the length of BC

43. the length of DC

44. the length of AD

45. the area of $\Delta ABC$

46. the area of $\Delta ADC$

47. In Figure 3.61, find the area of $\Delta ABC$.

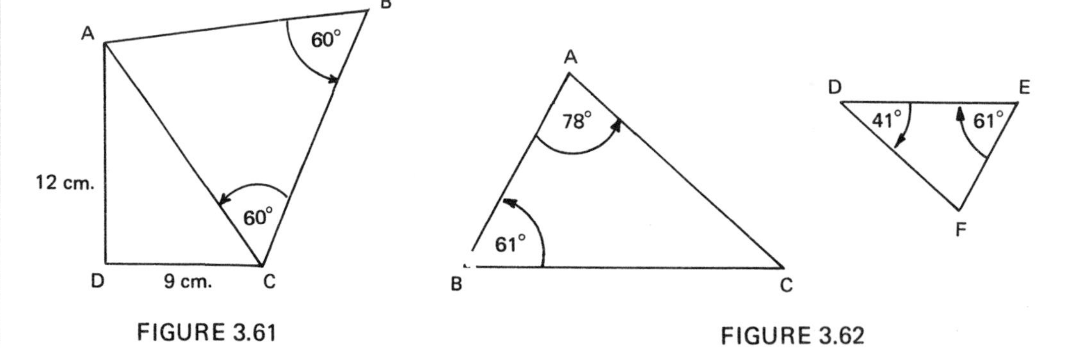


FIGURE 3.61

FIGURE 3.62

NAME

CLASS

ANSWERS

48. In Figure 3.62, show that $\triangle ABC$ and $\triangle DEF$ are similar.

48. ______

49. In Figure 3.63, show that $\triangle ABC$ and $\triangle DEF$ are similar.

49. ______

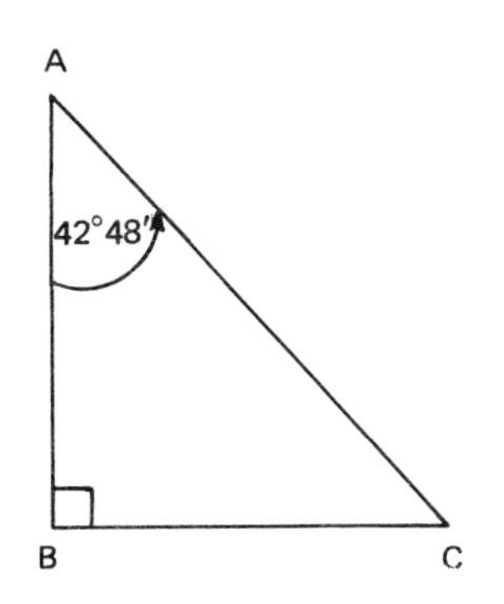


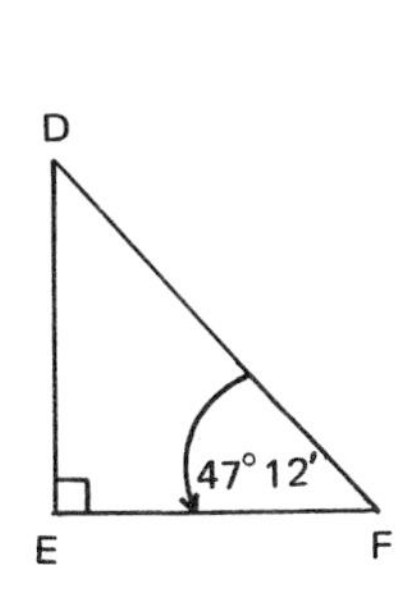


FIGURE 3.63

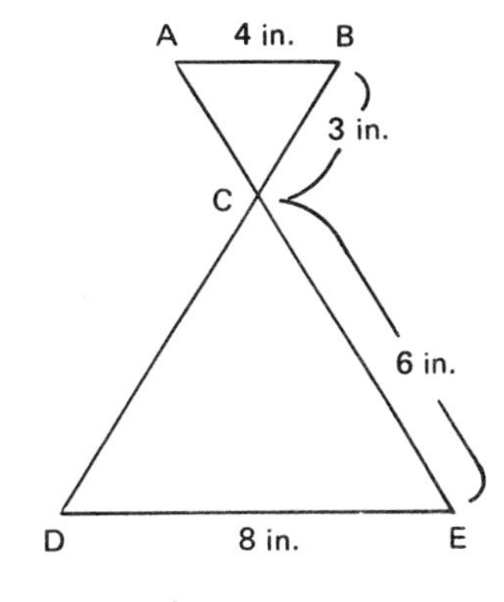


FIGURE 3.64

50. In Figure 3.64, suppose AB || DE.

(a) Show that $\triangle ABC$ and $\triangle CDE$ are similar.

(b) What is the ratio of similarity?

(c) Find the length of AC.

(d) Find the length of BD.

50. (a) ______

(b) ______

(c) ______

(d) ______

51. In Figure 3.65, suppose BC || DE.

(a) Show that $\triangle ABC$ and $\triangle ADE$ are similar.

(b) What is the ratio of similarity?

(c) Find the length of AC.

(d) Find the length of BD.

51. (a) ______

(b) ______

(c) ______

(d) ______

ANSWERS

52. (a)________________

(b)________________

53. ________________

54. ________________

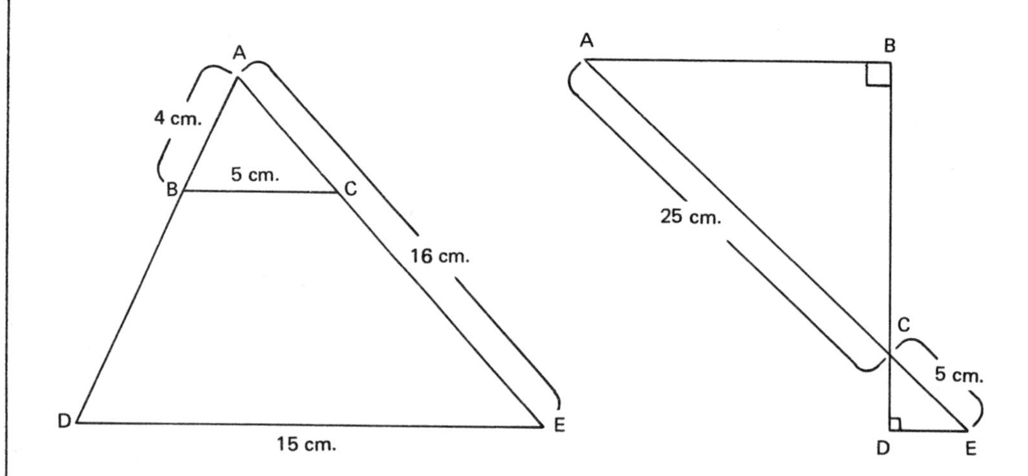


FIGURE 3.65

FIGURE 3.66

52. In Figure 3.66, suppose BD = 18 cm.

(a) Find the area of $\Delta CDE$.

(b) Show that $AB \parallel DE$.

53. A 16-foot ladder leaning against a wall makes a $30^\circ$-angle with the ground. (a) How far is the foot of the ladder from the bottom of the wall? (b) To 1 decimal digit, how high up the wall does the ladder reach?

54. As an airplane rises, it makes a $45^\circ$ angle with the runway. To 1 decimal digit, how many meters does the plane rise while traveling 800 meters in the air?

## 3.3 More on Circles

### BASIC FORMULAS

Consider a circle in which

$d$ = diameter

$r$ = radius

$C$ = circumference

and $A$ = area

Then recall that

$$d = 2r$$

$$C = \pi d = 2\pi r$$

and $$A = \pi r^2$$

### SECANTS, CHORDS, TANGENTS

A *secant* to a circle is a line that intersects the circle at two points, P and Q. In Figure 3.67, $\ell_1$ is a secant to the circle. The line segment PQ is known as a *chord.* A *tangent* to a circle is a line that intersects the circle at only one point. Thus, $\ell_2$ is tangent to the circle. *A tangent to a circle is perpendicular to the radius at the point of tangency.* In Figure 3.67, $OR \perp \ell_2$

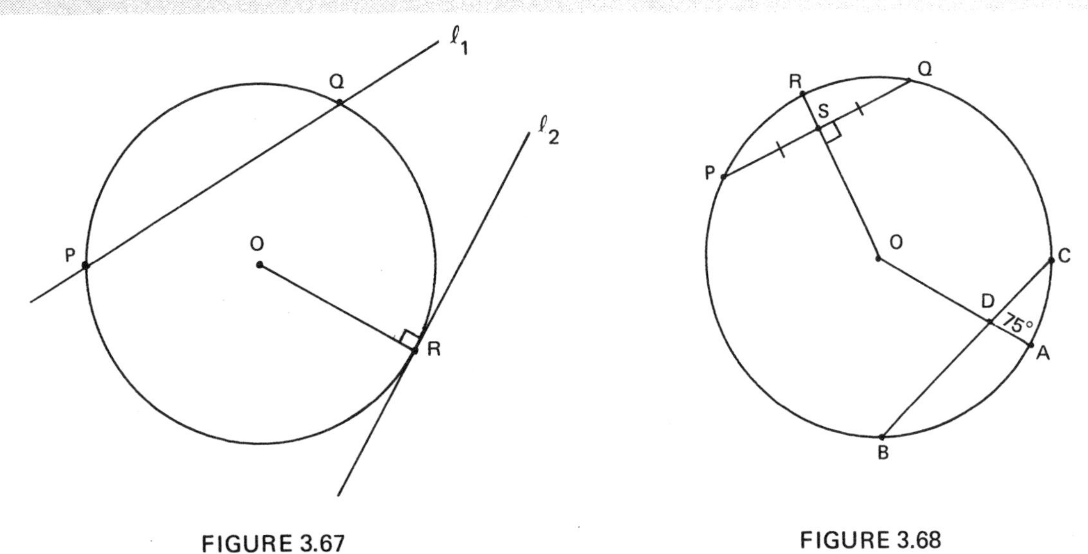


FIGURE 3.67

FIGURE 3.68

*A radius that bisects a chord is perpendicular to the chord. And a radius that intersects but does not bisect the chord is not perpendicular to the chord.* Thus, in Figure 3.68, $OR \perp PQ$ and PS = SQ. But OA is not perpendicular to BC, and BD is larger than DC.

EXAMPLE 1. In Figure 3.69, $\ell$ is tangent to the circle at S and O is the center of the circle.

(a) Show that $\angle$TSO is a right angle.

(b) Show that $\angle$ORP is a right angle.

SOLUTION. (a) The tangent, $\ell$, is perpendicular to the radius of the circle at the point of tangency. Thus, $\angle$TSO is a right angle.

(b) RQ = PQ − PR

= 8 cm. − 4 cm.

= 4 cm.

Thus, RQ = PR, and radius OA bisects chord PQ. Therefore, OA $\perp$ PQ, and $\angle$ORP is a right angle.

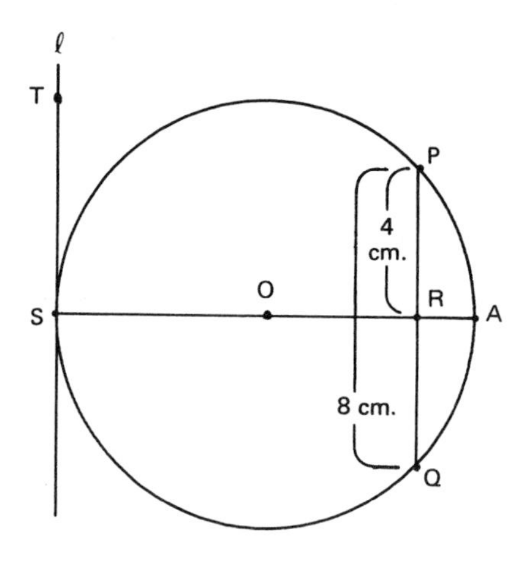


FIGURE 3.69

## CLASS EXERCISES

1. In Figure 3.70, $\ell$ is tangent to the circle at R, O is the center of the circle, and OP $\perp$ RS. Find the length of:

   (a) OP (b) RT (c) OT

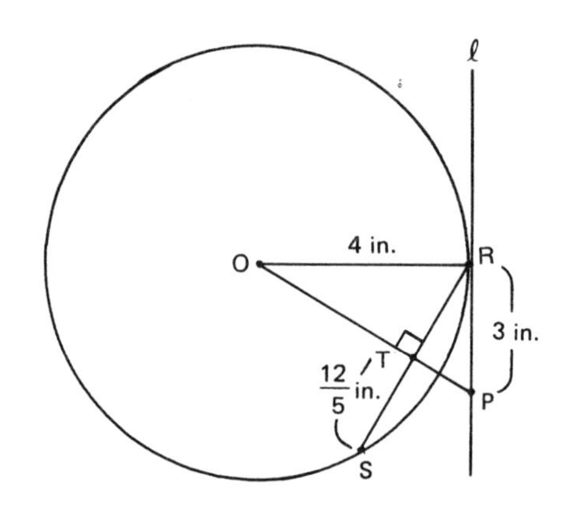


FIGURE 3.70

### CENTRAL ANGLES

Let A and B be two points on a circle whose center is at 0. $\angle$AOB is called a *central angle.* (See Figure 3.71.) The portion of the circle between A and B *in the interior of* $\angle$AOB is called the *arc intercepted by* $\angle$AOB, and is written $\widehat{AB}$. The arc $\widehat{AB}$ is also said to be intercepted by the chord AB and by the secant through A and B. Now consider *any* point C on the circle but not on arc $\widehat{AB}$. Then $\angle$ACB is called an *inscribed angle.* We also say that $\angle$ACB intercepts $\widehat{AB}$. *When a central angle,* $\angle$AOB, *and an inscribed angle,* $\angle$ACB, *intercept the same arc, then*

$$\angle AOB = 2\,\angle ACB$$

In Figure 3.71, note that $\angle$ADB, an inscribed angle, also intercepts arc $\widehat{AB}$. Thus,

$$\angle AOB = 2\ \angle ADB = \angle ACB$$

and therefore,

$$\angle ADB = \angle ACB$$

Similarly, $\angle AEB$ is an inscribed angle that intercepts arc $\overset{\frown}{AB}$.

$$\angle AEB = \angle ACB$$

*Any two inscribed angles that intercept the same arc have equal measures.*

EXAMPLE 2. In Figure 3.72, O is the center of the circle. Find the measure of:

(a) $\angle AOB$ (b) $\angle ADB$

SOLUTION. (a) $\angle AOB$ is a central angle that intercepts the same arc $\overset{\frown}{AB}$, as does $\angle ACB$, an inscribed angle. Therefore,

$$\begin{aligned}\angle AOB &= 2\ \angle ACB \\ &= 2 \cdot 70^\circ \\ &= 140^\circ\end{aligned}$$

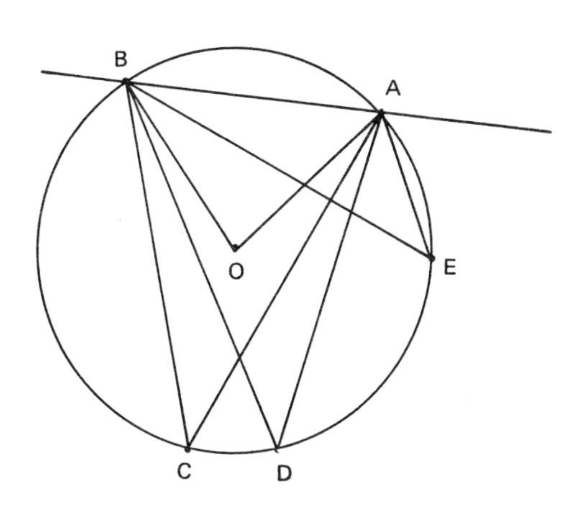


FIGURE 3.71 $\angle AOB = 2\angle ACB$

FIGURE 3.72

(b) $\angle ADB$ is another inscribed angle that intercepts $\overset{\frown}{AB}$. Therefore,

$$\angle ADB = \angle ACB = 70^\circ$$

A central angle of $360^\circ$ intercepts the entire circle. A central angle of $90^\circ$ intercepts an arc whose length is ¼ of the circumference. (See Figure 3.73.) Note that

$$\frac{90^\circ}{360^\circ} = \frac{1}{4}$$

In general, the following proportion holds:

$$\frac{\text{central angle}}{360^\circ} = \frac{\text{arc length}}{\text{circumference}}$$

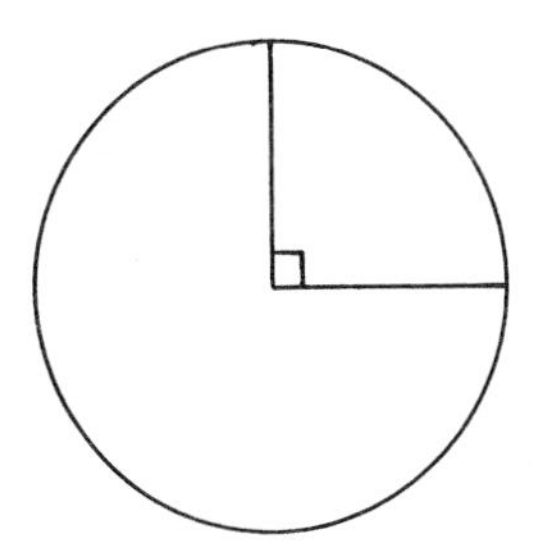

FIGURE 3.73

EXAMPLE 3. A circle has radius 9 inches. Determine the length of an arc intercepted by a central angle of $60^\circ$.

SOLUTION. Let C = circumference, r = radius. Then

$$C = 2\pi r$$

$$C = 2\pi \times 9 \text{ in.}$$

$$C = 18\pi \text{ in.}$$

$$\frac{\text{central angle}}{360^\circ} = \frac{\text{arc length}}{\text{circumference}}$$

$$\frac{60^\circ}{360^\circ} = \frac{\text{arc length}}{18\pi \text{ in.}}$$

$$\frac{1}{6} = \frac{\text{arc length}}{18\pi \text{ in.}}$$

$$\frac{18\pi \text{ in.}}{6} = \text{arc length}$$

$$3\pi \text{ in.} = \text{arc length}$$

Note that $3\pi$ ($\approx 3 \times 3.14$) is slightly more than 9. Thus, in Figure 3.74, the arc intercepted is slightly longer than the radius of the circle.

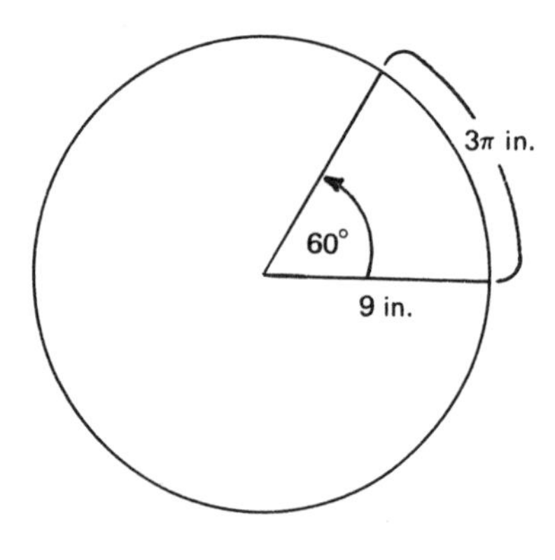

FIGURE 3.74

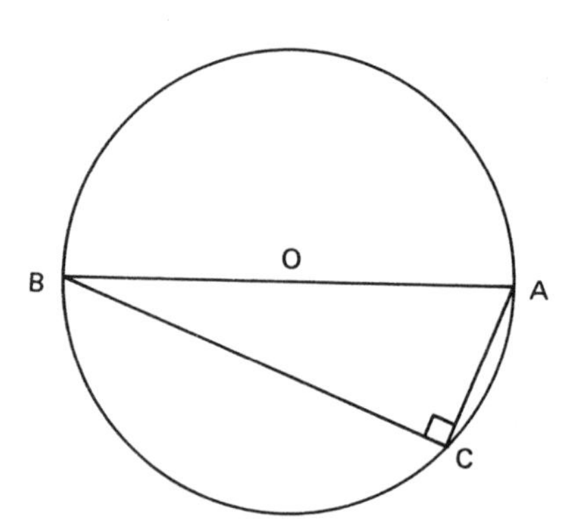

FIGURE 3.75
∠ AOB is a straight angle.
$\overset{\frown}{AB}$ is a semicircle.
∠ACB is a right angle.

A central angle of 180° (that is, a straight angle) intercepts an arc whose length is one-half the circumference. Such an arc is known as a *semicircle.* Observe, in Figure 3.75, that ***a chord that intercepts a semicircle is a diameter. And an inscribed angle that intercepts a semicircle is a right angle*** **(½ × 180°).**

The portion of a circular region determined by central angle ∠AOB is known as *sector* AOB. In Figure 3.76, the sector determined by a right angle (90°) is indicated. It appears that the area of the sector is one-fourth that of the circle. As previously noted,

$$\frac{90^\circ}{360^\circ} = \frac{1}{4}$$

In general,

$$\frac{\text{central angle}}{360^\circ} = \frac{\text{area of sector}}{\text{area of circle}}$$

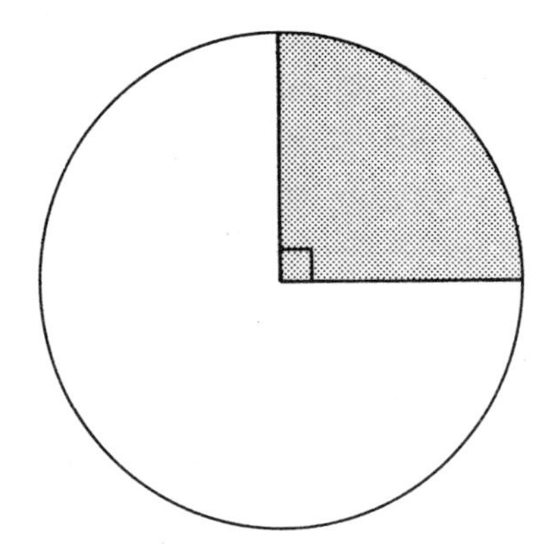

FIGURE 3.76
The sector is shaded.

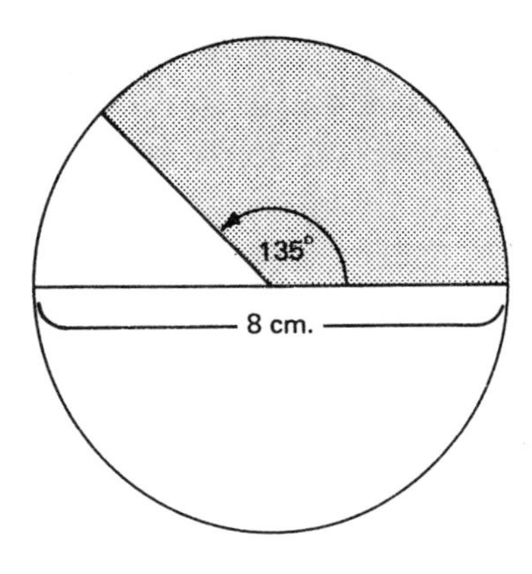


FIGURE 3.77
The circle has area $16\pi$ cm.$^2$.
The (shaded) sector has area $6\pi$ cm.$^2$.

EXAMPLE 4. The diameter of a circle is 8 centimeters. Find the area of a sector of the circle determined by a 135°-central angle. (See Figure 3.77.)

SOLUTION. Let r = radius, d = diameter, A = area of circle.

$$r = \frac{1}{2} \cdot d$$

$$= \frac{1}{2} \times 8 \text{ cm.}$$

$$= 4 \text{ cm.}$$

$$A = \pi \times (4 \text{ cm.})^2$$

$$= 16\pi \text{ cm.}^2$$

$$\frac{\text{central angle}}{360^\circ} = \frac{\text{area of sector}}{\text{area of circle}}$$

$$\frac{135^\circ}{360^\circ} = \frac{\text{area of sector}}{16\pi \text{ cm.}^2}$$

$$\frac{3}{8} = \frac{\text{area of sector}}{16\pi \text{ cm.}^2}$$

$$\frac{3}{8} \times 16\pi \text{ cm.}^2 = \text{area of sector}$$

$$6\pi \text{ cm.}^2 = \text{area of sector}$$

## CLASS EXERCISES

2. In Figure 3.78, O is the center of the circle. Find:

   (a) the measure of $\angle$ ACB

   (b) the measure of $\angle$ OBA

   (c) the length of $\widehat{AB}$

   (d) the area of sector AOB

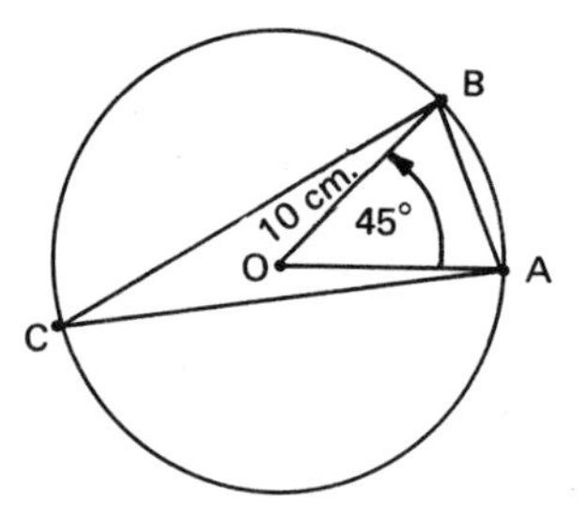


FIGURE 3.78

## SOLUTIONS TO CLASS EXERCISES

1. (a)

$$OR \perp \ell$$

Therefore, $\triangle$ORP is a right triangle with hypotenuse OP. By the Pythagorean Theorem, OP has length 5 inches.

(b)

$$OP \perp RS$$

Therefore, OP bisects RS. Thus, RT has length $\frac{12}{5}$ inches.

(c) $\triangle$OTR is a right triangle.

$$(OT)^2 + (\tfrac{12}{5}\text{ in.})^2 = (4\text{ in.})^2$$

$$(OT)^2 + \frac{144}{25}\text{ in.}^2 = 16\text{ in.}^2$$

$$(OT)^2 = \frac{256}{25}\text{ in.}^2$$

$$OT = \frac{16}{5}\text{ in.}$$

2. (a) $\angle ACB = \frac{1}{2} \cdot 45° = 22.5°$ (or $22°30'$)

   (b) OB and OA are radii. Therefore, $\triangle$OBA is isosceles. Thus,

$$\angle OBA = \angle BAO$$

$$2\angle OBA = 180° - 45°$$

$$2\angle OBA = 135°$$

$$\angle OBA = 67.5°$$

(c)

$$\frac{45^\circ}{360^\circ} = \frac{\text{length } \widehat{AB}}{\text{circumference}}$$

$$\frac{1}{8} = \frac{\text{length } \widehat{AB}}{20\pi \text{ cm.}}$$

$$\frac{1}{8} \times 20\pi \text{ cm.} = \text{length } \widehat{AB}$$

$$2.5\pi \text{ cm.} = \text{length } \widehat{AB}$$

(d)

$$\frac{45^\circ}{360^\circ} = \frac{\text{area of sector}}{\text{area of circle}}$$

$$\frac{1}{8} = \frac{\text{area of sector}}{100\pi \text{ cm.}^2}$$

$$\frac{1}{8} \times 100\pi \text{ cm.}^2 = \text{area of sector}$$

$$12.5\pi \text{ cm.}^2 = \text{area of sector}$$

NAME ______

CLASS ______

ANSWERS

## HOME EXERCISES

Throughout these exercises, assume O is the center of each circle.

1. Refer to Figure 3.79. $\angle BOD$ is a straight angle. Match each entry in column (a) with its description in column (b).

| Column (a) | Column (b) |
|---|---|
| $\overset{\frown}{AB}$ | chord |
| $\angle AOB$ | semicircle |
| $\angle ACB$ | an arc that is not a semicircle |
| BC | tangent |
| $\ell_1$ | diameter |
| $\ell_2$ | secant |
| BD | central angle |
| $\overset{\frown}{BD}$ | inscribed angle |

1.

$\overset{\frown}{AB}$ ______

$\angle AOB$ ______

$\angle ACB$ ______

BC ______

$\ell_1$ ______

$\ell_2$ ______

BD ______

$\overset{\frown}{BD}$ ______

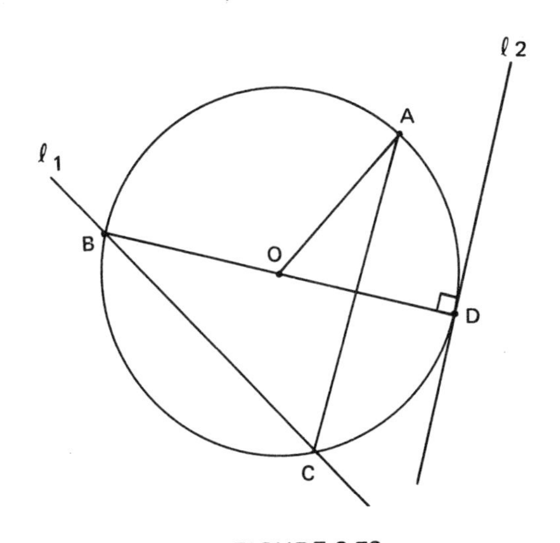


FIGURE 3.79

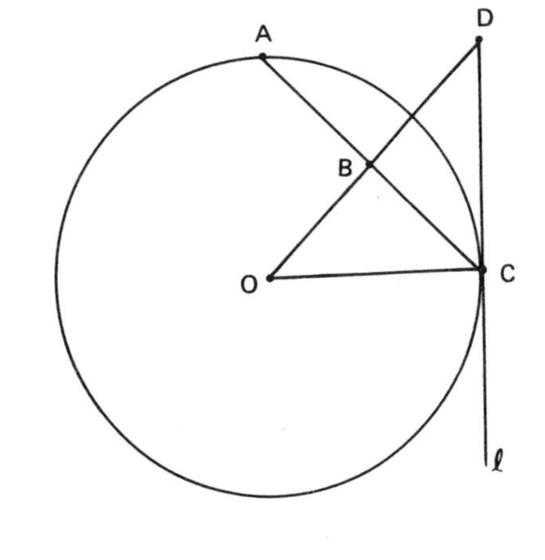


FIGURE 3.80

Exercises 2-9 refer to Figure 3.80. $\ell$ is tangent to the circle at C. The radius of the circle is 5 centimeters. $AB = BC = \frac{5\sqrt{2}}{2}$ cm. Find:

2. the measure of $\angle OBC$

3. the length of OB

4. the area of $\triangle OBC$

5. the measure of $\angle DCO$

6. the length of BD

7. the length of OD

2. ______

3. ______

4. ______

5. ______

6. ______

7. ______

ANSWERS

8. ____________

9. ____________

10. (a) ____________

(b) ____________

(c) ____________

11. (a) ____________

(b) ____________

12. (a) ____________

(b) ____________

(c) ____________

8. the area of $\triangle OCD$ 9. the area of $\triangle BCD$

10. In Figure 3.81, find:

(a) the measure of $\angle ACB$ (b) the length of $\overset{\frown}{AB}$ (c) the area of sector AOB

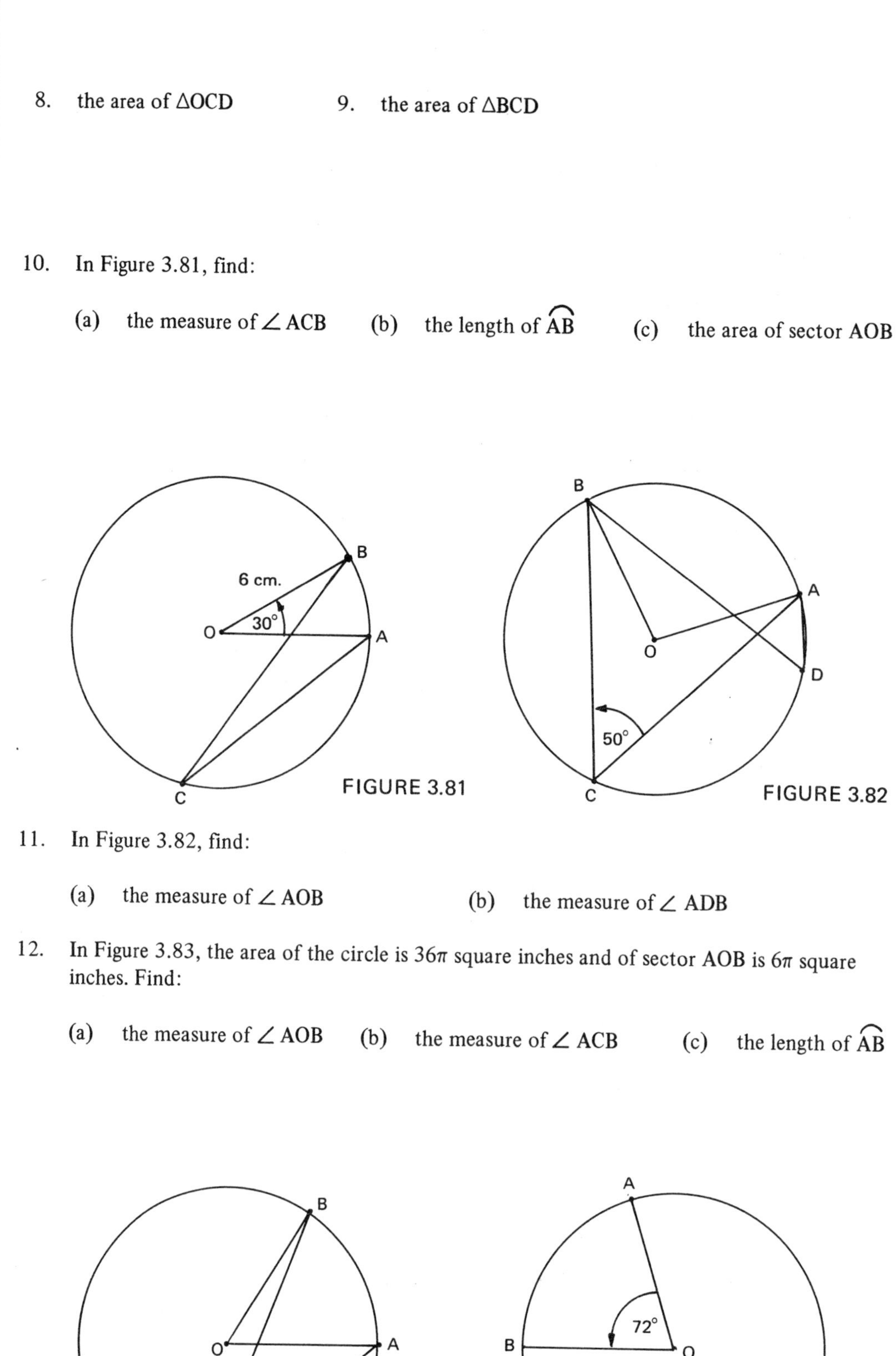


FIGURE 3.81

FIGURE 3.82

11. In Figure 3.82, find:

(a) the measure of $\angle AOB$ (b) the measure of $\angle ADB$

12. In Figure 3.83, the area of the circle is $36\pi$ square inches and of sector AOB is $6\pi$ square inches. Find:

(a) the measure of $\angle AOB$ (b) the measure of $\angle ACB$ (c) the length of $\overset{\frown}{AB}$



FIGURE 3.83

FIGURE 3.84

NAME

CLASS

ANSWERS

13. In Figure 3.84, $\widehat{AB}$ is of length $2\pi$ inches. Find:

(a) the circumference of the circle (b) the radius (c) the area of sector AOB

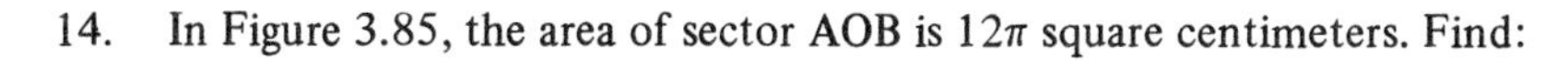

14. In Figure 3.85, the area of sector AOB is $12\pi$ square centimeters. Find:

(a) the area of the circle (b) the radius (c) the length of $\widehat{AB}$

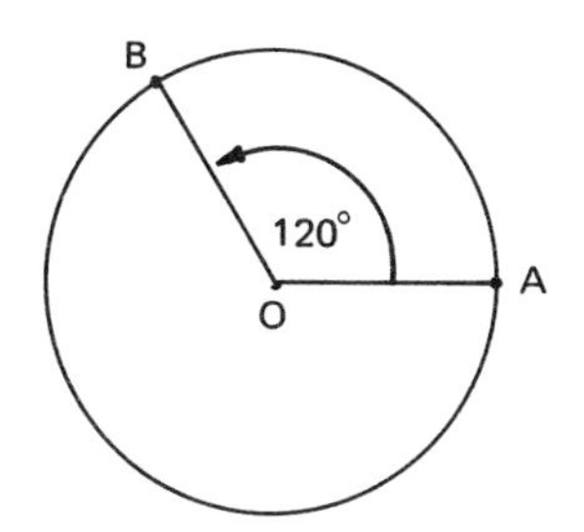


FIGURE 3.85

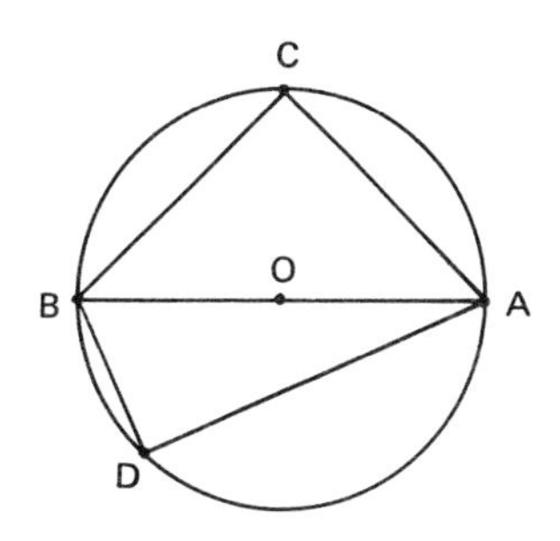


FIGURE 3.86

15. In Figure 3.86, suppose that AB is a diameter of the circle. Find:

(a) the measure of $\angle$ BCA (b) the measure of $\angle$ADB

For exercises 16-21, find:

(a) the length of the boundary (b) the area of each shaded region

16.

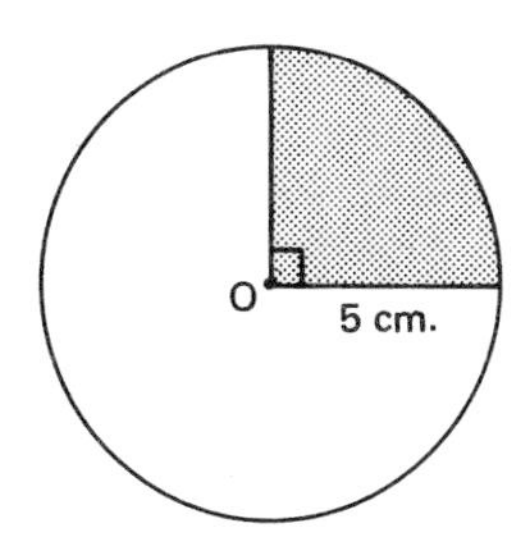


FIGURE 3.87

17.

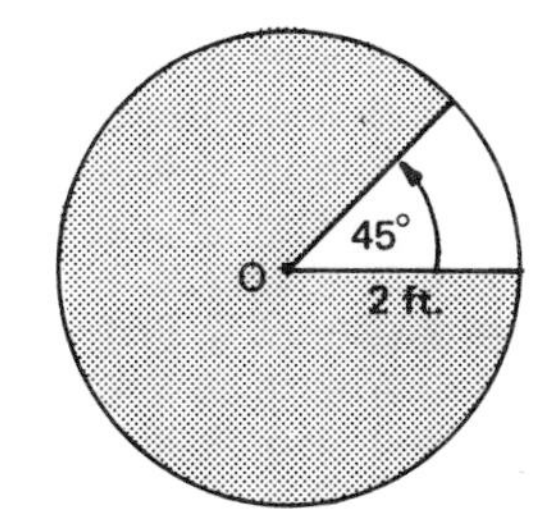


FIGURE 3.88

18.

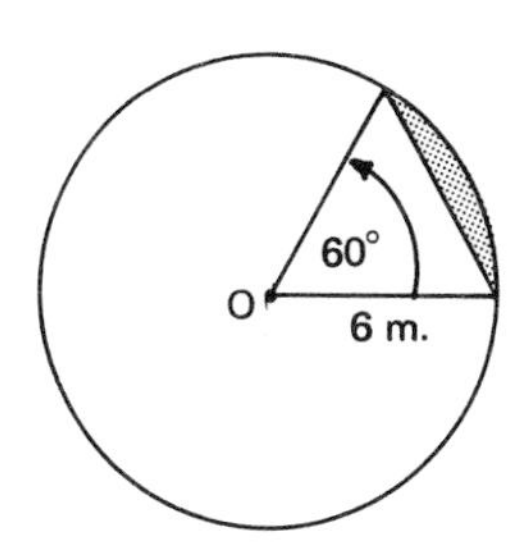


FIGURE 3.89

13. (a) ________

(b) ________

(c) ________

14. (a) ________

(b) ________

(c) ________

15. (a) ________

(b) ________

16. (a) ________

(b) ________

17. (a) ________

(b) ________

18. (a) ________

(b) ________

ANSWERS

19. (a) ______________

(b) ______________

20. (a) ______________

(b) ______________

21. (a) ______________

(b) ______________

22. ______________

23. ______________

24. ______________

25. ______________

26. ______________

27. ______________

28. ______________

29. ______________

30. ______________

31. ______________

19. 20. 21.

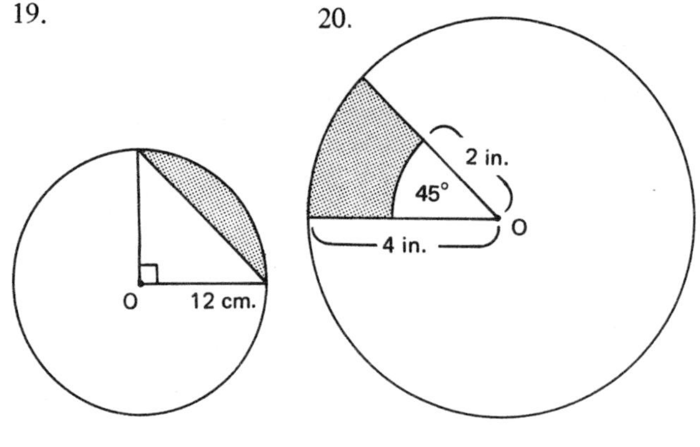


FIGURE 3.90 FIGURE 3.91

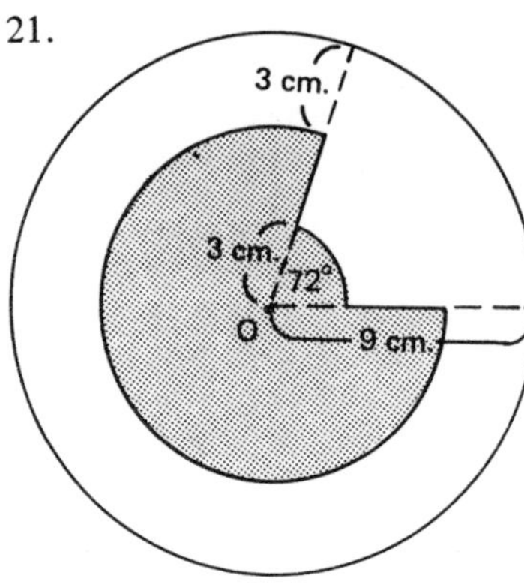


FIGURE 3.92

Exercises 22-25 refer to Figure 3.93. Find:

22. the area of $\triangle AOB$

23. the length of $\overset{\frown}{BC}$

24. the measure of $\angle$ CBO

25. the measure of $\angle$ ADC

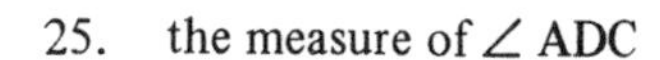

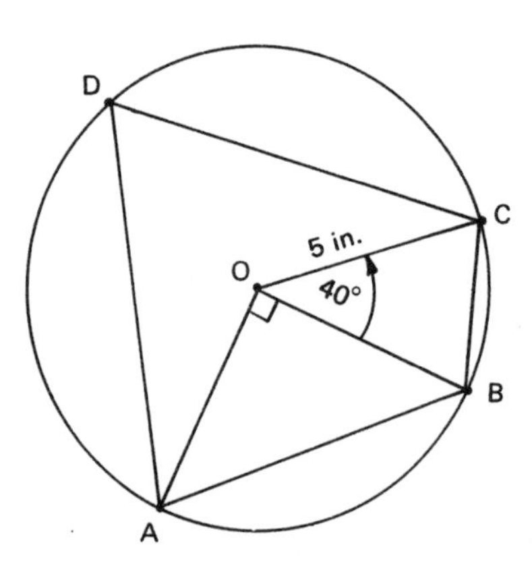


FIGURE 3.93

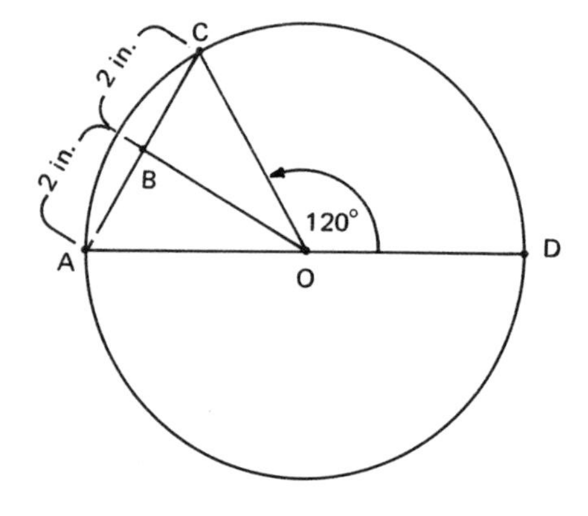


FIGURE 3.94

Exercises 26-31 refer to Figure 3.94. Find:

26. the measure of $\angle$ ABO

27. the measure of $\angle$ OAC

28. the measure of $\angle$ AOB

29. the length of AO

30. the area of $\triangle BOA$

31. the area of sector OCD

*Let's Review Chapter 3.*

3.1 Angles

1. Convert 28.6° into degrees and minutes.

2. Write 66°12′ in decimal notation.

3. Classify as either a right angle, a straight angle, an acute angle, or an obtuse angle.

   (a) 62° (b) 145° (c) 180°

4. Find the measure of (a) the complement, (b) the supplement of an angle of 52°49′.

5. An angle measures 20° more than its supplement. Find the measure of the angle.

6. In Figure 3.95, assume $\ell_1 \parallel \ell_2$. Find the measure of: (a) $\angle 1$ (b) $\angle 2$

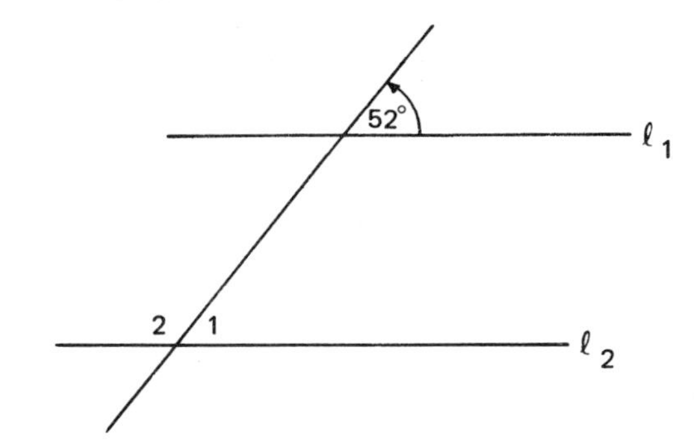


FIGURE 3.95

FIGURE 3.96

3.2 More on Triangles

7. One acute angle of a right triangle measures 43°15′. Find the measure of the other acute angle.

8. In Figure 3.96, AB = AC and AD is the height corresponding to side BC. (a) Find the length of BC. (b) Find the area of ΔABC.

9. Find the area of a 30°, 60°, 90°-triangle whose hypotenuse is 10 inches.

10. Find (a) the height and (b) the area of an equilateral triangle whose side length is 5 centimeters.

3.3 More on Circles

Refer to Figure 3.97. O is the center of the circle. Find:

11. the measure of ∠ACB

12. the measure of ∠ADB

13. the length of arc $\widehat{AB}$

14. the area of sector AOB

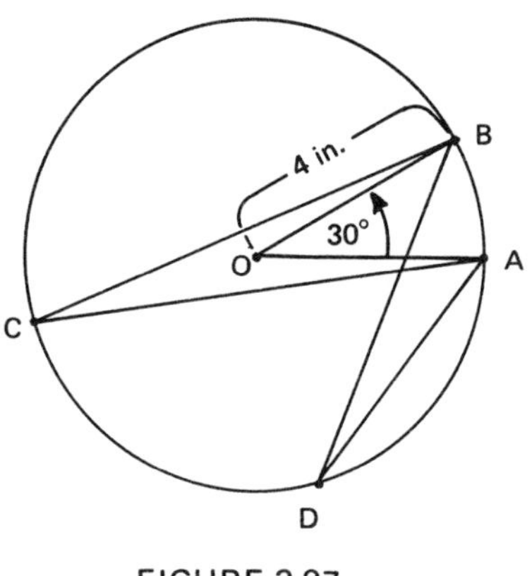

FIGURE 3.97

*Try These Exam Questions for Practice.*

1. Convert 44.75° into degrees and minutes.

2. The complement of angle A measures one-fourth as much as the supplement of angle A. Find angle A.

3. Find (a) the height and (b) the area of an equilateral triangle with side length 8 inches.

In Figure 3.98, O is the center and diameter AC measures 6 inches. Find:

4. the measure of ∠AEC

5. the length of arc $\widehat{AB}$

6. the area of sector AOB

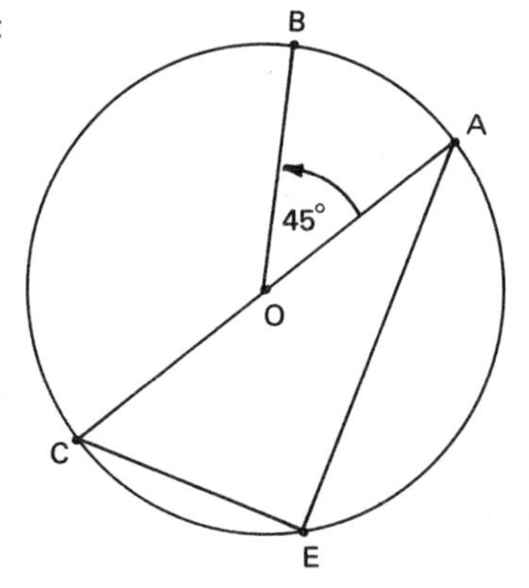

FIGURE 3.98

# ANSWERS

## Chapter 1

Diagnostic Test, p. 1

1.1 (a) 84 inches (b) .06 meter (c) 25 miles

1.2 (a) 14 centimeters (b) 25 inches (c) 35 sq. in.

1.3 (a) 15 sq. cm. (b) 4 sq. ft. (c) 300 sq. cm.

1.4 (a) $9\pi$ sq. cm. (b) $36\pi$ sq. in.

1.5 (a) 2 inches (b) 30 inches (c) 3 centimeters

Home Exercises

1.1, p. 9

1. 8 feet 3. 6 yards 5. 7 miles 7. $\frac{8}{9}$ yard
9. 4.5 yards 11. 15,840 ft. 13. 28 feet 15. 528 feet
17. 5000 meters 19. 4.05 meters 21. 2.44 kilometers
23. 358 centimeters 25. 1969 inches 27. 25 centimeters
29. 31 miles 31. 48 kilometers 33. 338 kilometers
35. (a) 3600 inches (b) 9144 centimeters
37. 2 gallons

1.2, p. 19

1. 15 in. 3. 38 in. 5. 48 cm. 7. 22 in.
9. 20 feet 11. 4 feet 13. 6 miles 15. 10 inches
17. 280 feet 19. 360 feet 21. 12 sq. in. 23. 300 sq. cm.
25. $\frac{3}{2}$ sq. in. 27. 4 inches 29. 3 cm. 31. 6 yards
33. 81 square inches 35. 24 square feet 37. 5333 square yards
39. $28 41. 15 inches by 20 inches
43. 18 feet by 20 feet 45. 100 square inches

1.3, p. 35

1. 8 sq. in. 3. $\frac{9}{4}$ sq. in. 5. 42 sq. in. 7. 20 inches
9. 3 inches 11. 50 yards 13. 20 cm. 15. 12 inches
17. $\sqrt{15}$ cm. 19. 12 feet 21. 80 sq. in. 23. 120 sq. cm.
25. 140 sq. in. 27. 9 inches 29. 8 inches 31. 25 sq. in.
33. 3 sq. in. 35. 42 in.$^2$ 37. 11 inches 39. $\frac{70}{9}$ inches
41. 4 inches 43. 52 cm.$^2$ 45. 180 cm.$^2$ 47. 180 cm.$^2$

1.4, p. 47

1. 20 inches
3. 24 cm.
5. 2 inches
7. 12 cm.
9. (a) $\pi$ inches (b) 3 inches
11. (a) $9\pi$ yards (b) 28 yards
13. (a) $2\pi$ feet (b) 6 feet
15. (a) $\pi$ miles (b) 3 miles
17. (a) $25\pi$ dm. (b) 79 dm.
19. (a) $9\pi$ sq. yd. (b) 28 sq. yd.
21. (a) $400\pi$ square inches (b) 1257 square inches
23. (a) $36\pi$ square dekameters (b) 113 square dekameters
25. (a) $10{,}000\pi$ square miles (b) 31,416 square miles
27. (a) $576\pi$ sq. yd. (b) 1810 sq. yd.
29. $9\pi$ sq. in.
31. (a) 10 cm. (b) 5 cm. (c) $25\pi$ sq. cm.
33. 188 ft.
35. (a) $21\pi$ in.$^2$ (b) $(100 - 25\pi)$ in.$^2$ (c) $(64-8\pi)$ cm.$^2$ (d) $(36 + \frac{9\pi}{2})$ in.$^2$, or $\frac{72 + 9\pi}{2}$ in.$^2$

1.5, p. 63

1. similar
3. similar
5. (a) 1 in. (b) 3 (c) 9
7. (a) 2 cm. (b) 2 (c) 4
9. 3 inches
11. (a) 4 (b) 16
13. 12 inches
15. (a) 3 (b) 9
17. 2
19. (a) 20 in. (b) 3 (c) 63 in. (d) 36 in. (e) 9
21. (a) 12 cm. (b) 3 (c) $\frac{10}{3}$ cm. (d) 9
23. (a) $\frac{5}{2}$ in. (b) 2 (c) 8 in. (d) $\frac{5}{2}$ in. (e) 3 in. (f) 4
25. 5 cm.
27. 24 mm.
29. 3 km.
31. 6 inches

Let's Review Chapter 1, p. 69

1. 42 inches
2. 567 cm.
3. 393.7 in.
4. 52 km.
5. 18 cm.
6. 36 inches
7. 63 sq. cm.
8. 16 sq. dm.
9. 7 inches
10. 108 in.$^2$
11. 45 cm.$^2$
12. 42 in.$^2$
13. 12 meters
14. 54 in.$^2$
15. $\frac{9}{2}$ cm.
16. (a) $4\pi$ sq. ft. (b) 12.6 sq. ft.
17. (a) 8 cm. (b) 4 cm. (c) $16\pi$ sq. cm.
18. (a) 3 cm. (b) 2 (c) 4
19. 36 sq. ft.
20. 3 inches

Try These Exam Questions for Practice, p. 70

1. (a) 18 cm. (b) 18 cm.$^2$
2. 12 feet
3. 28 sq. in.
4. (a) $6\pi$ cm. (b) 3 cm. (c) $9\pi$ cm.$^2$
5. 108 in.$^2$

## Chapter 2

Diagnostic Test, p. 73

2.1 (a) 6 cubic feet (b) 22 sq. ft. (c) 9 feet

2.2 (a) volume = $\frac{500\pi}{3}$ cubic inches, surface area = $100\pi$ sq. in.

(b) volume = $45\pi$ cc., lateral surface area = $30\pi$ sq. cm. (c) $324\pi$ cubic inches

Home Exercises

2.1, p. 85

1. (a) 20 in.$^3$ (b) 48 in.$^2$
3. (a) 180 cc. (b) 222 sq. cm.
5. (a) 800 cc. (b) 600 cm.$^2$
7. (a) 8 cc. (b) 24 cm.$^2$
9. (a) 216 cc. (b) 216 cm.$^2$
11. 5 inches
13. 4 inches
15. 1260 ft.$^3$
17. (a) 198 in.$^3$ (b) 234 in.$^2$
19. similar
21. not similar
23. (a) 10 cm. (b) 5 (c) 2 cm. (d) 25 (e) 125
25. 2160 cubic feet
27. 1 ft. and $\frac{1}{2}$ ft. (or 6 inches)
29. 1000 cubic inches

2.2, p. 105

1. (a) $\frac{256\pi}{3}$ km.$^3$, (b) 268 km.$^3$ (c) $64\pi$ km.$^2$
3. (a) $972\pi$ m.$^3$ (b) 3054 m.$^3$ (c) $324\pi$ m.$^2$
5. (a) $\frac{\pi}{6}$ cubic inch (b) .5 cubic inch (c) $\pi$ sq. in.
7. 2 yards
9. $\frac{256\pi}{3}$ cubic millimeters
11. (a) $160\pi$ cubic inches (b) 503 sq. in. (c) $80\pi$ sq. in.
13. (a) $54\pi$ in.$^3$ (b) 170 in.$^3$ (c) $36\pi$ in.$^2$
15. 6 inches
17. 10 inches
19. (a) $30\pi$ cubic inches (b) 94 cubic inches (c) $3\sqrt{109}\cdot\pi$ in.$^2$
21. (a) $4\pi$ cubic feet (b) 13 cubic feet (c) $2\pi\sqrt{13}$ sq. ft.
23. 6 in.
25. 3 in.
27. $100\pi$ in.$^3$
29. $\frac{500\pi}{3}$ in.$^3$
31. $\frac{112\pi}{3}$ in.$^3$
33. (a) 2 in. (b) 4 (c) 16 (d) 64
35. 2
37. 15 inches
39. 2 inches

Let's Review Chapter 2, p. 111

1. (a) 60 in.$^3$ (b) 94 in.$^2$
2. (a) 64 cc. (b) 96 cm.$^2$
3. (a) 4 cm. (b) 2 (c) 3 cm. (d) 4 (e) 8
4. (a) $\frac{256\pi}{3}$ in.$^3$ (b) 268 in.$^3$ (c) $64\pi$ in.$^2$
5. $36\pi$ cc.
6. (a) $36\pi$ in.$^3$ (b) 113 in.$^3$ (c) $24\pi$ in.$^2$
7. 2 cm.
8. (a) $32\pi$ in.$^3$ (b) 100 in.$^3$ (c) $8\sqrt{13}\cdot\pi$ in.$^2$

Try These Exam Questions for Practice, p. 113

1. (a) 450 in.$^3$ (b) 370 sq. in.
2. 10 cm.
3. (a) $\frac{4000\pi}{3}$ cubic inches (b) $400\pi$ square inches
4. $160\pi$ cubic inches
5. 301 cubic centimeters

## Chapter 3

Diagnostic Test, p. 115

3.1 (a) 25°24′ (b) 53°38′ (c) 41°

3.2 (a) 98° (b) $\frac{9\sqrt{3}}{2}$ sq. cm. (c) 10 inches

3.3 (a) 30° (b) $\frac{5\pi}{3}$ cm. (c) $\frac{25\pi}{6}$ cm.$^2$

Home Exercises

3.1, p. 127

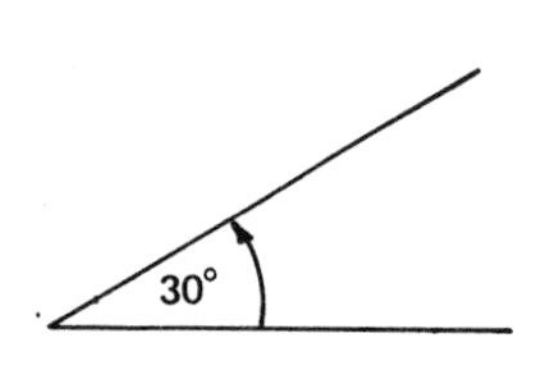


1. FIGURE 3A

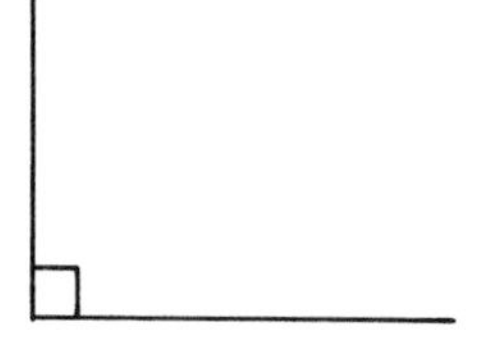

3. FIGURE 3B

| | | | | | | | |
|---|---|---|---|---|---|---|---|
| 5. | 12′ | 7. | 19°30′ | 9. | 100°48′ | 11. | 5.5° |
| 13. | 39.8° | 15. | 80.05° | 17. | acute | 19. | obtuse |
| 21. | 80° | 23. | 45° | 25. | 70.8° | 27. | 130° |
| 29. | 142.6° | 31. | 179°21′ | 33. | 60° | 35. | 80° |
| 37. | 45° | 39. | 22.5° | 41. | 84°38′ | 43. | 84°38′ |
| 45. | 70° | 47. | 70° | 49. | 110° | 51. | 88° |
| 53. | 92° | 55. | 63° | 57. | 8 inches | 59. | 12 inches |

3.2, p. 143

1. 70°
3. 82°10′
5. 26.4°
7. 45°
9. 42°48′
11. 60° and 30°
13. 75°, 55°, 50°
15. (a), (c), (d)
17. (a) 4 cm. (b) 80 cm.$^2$
19. (a) $8\sqrt{2}$ cm. (b) 32 cm.$^2$
21. $a = 3\sqrt{3}$ in., $b = 3$ in.
23. $b = \sqrt{3}$ feet, $c = 2\sqrt{3}$ feet
25. (a) $\frac{7\sqrt{3}}{2}$ inches (b) 21 inches (c) $\frac{49\sqrt{3}}{4}$ sq. in.
27. (a) 6 inches (b) $3\sqrt{3}$ inches
29. 45°
31. 105°
33. $2\sqrt{2}$ in.
35. $(2 + 2\sqrt{3})$ in.
37. 2 in.$^2$
39. $2\sqrt{3}$ in.$^2$
41. 30°

43. 1.5 cm.

45. $\frac{9\sqrt{3}}{4}$ cm.$^2$

47. $\frac{225\sqrt{3}}{4}$ cm.$^2$

49. $\angle A = \angle D = 42°48'$, $\angle B = \angle E = 90°$, $\angle C = \angle F = 47°12'$

51. (a) Both triangles share $\angle$ A. Also, $\angle$CBA = $\angle$EDA and $\angle$ACB = $\angle$AED.
(b) 3 (c) $\frac{16}{3}$ cm. (d) 8 cm.

53. (a) 8 feet (b) 13.9 feet

3.3, p. 159

1.

| column (a) | column (b) |
| --- | --- |
| $\widehat{AB}$ | an arc that is not a semicircle |
| $\angle$AOB | central angle |
| $\angle$ACB | inscribed angle |
| BC | chord |
| $\ell_1$ | secant |
| $\ell_2$ | tangent |
| BD | diameter |
| $\widehat{BD}$ | semicircle |

3. $\frac{5\sqrt{2}}{2}$ cm.

5. 90°

7. $\frac{5\sqrt{2}}{2}$ cm.

9. $\frac{25}{4}$ cm.$^2$

11. (a) 100° (b) 50°

13. (a) $10\pi$ in. (b) 5 in. (c) $5\pi$ in.$^2$

15. (a) 90° (b) 90°

17. (a) $\frac{8+7\pi}{2}$ ft. (b) $\frac{7\pi}{2}$ ft.$^2$

19. (a) $(12\sqrt{2} + 6\pi)$ cm. (b) $(36\pi - 72)$ cm.$^2$

21. (a) $\frac{30+54\pi}{5}$ cm. (b) $\frac{153\pi}{5}$ cm.$^2$

23. $\frac{10\pi}{9}$ in.

25. 65°

27. 60°

29. 4 in.

31. $\frac{16\pi}{3}$ in.$^2$

Let's Review Chapter 3, p. 163

1. 28°36′

2. 66.2°

3. (a) acute (b) obtuse (c) straight

4. (a) 37°11′ (b) 127°11′

5. 100°

6. (a) 52° (b) 128°

7. 46°45′

8. (a) 8 inches (b) 64 sq. in.

9. $\frac{25\sqrt{3}}{2}$ square inches

10. (a) $\frac{5\sqrt{3}}{2}$ cm. (b) $\frac{25\sqrt{3}}{4}$ sq. cm.

11. 15°

12. 15°

13. $\frac{2\pi}{3}$ in.

14. $\frac{4\pi}{3}$ in.$^2$

Try These Exam Questions for Practice, p. 164

1. 44°45′

2. 60°

3. (a) $4\sqrt{3}$ in. (b) $16\sqrt{3}$ sq. in.

4. 90°

5. $\frac{3\pi}{4}$ in.

6. $\frac{9\pi}{8}$ sq. in.